新楼盘

NEWHOUSE 图解地产与设计

43

商业综合体

中国林业出版社

上海中建建筑設計院有限公司

上海中建建筑设计院有限公司，成立于1984年，隶属于中国建筑工程总公司，是国家批准的拥有建筑工程设计甲级、装饰设计甲级和风景园林设计乙级、城市规划编制乙级资质的综合性设计公司。公司是中国勘察设计协会和上海市勘察设计协会会员单位，并入选为中国2010年上海世博会建筑设计类九家推荐服务供应商之一。2011年，公司荣获中国勘察设计协会首批颁发的"诚信单位"荣誉称号。

公司长期面向海内外开展设计业务，先后与美国、加拿大、法国、比利时、香港等国家和地区的著名设计公司合作设计了多项工程，设计作品遍及国内二十多个省市自治区及海外的俄罗斯、阿尔及利亚等国家，有多项设计作品荣获建设部、上海市及中国建筑工程总公司的各类奖项。

公司将本着敬业、诚信、协作、创新的企业精神，恪守诚信原则，聚焦客户需求，为客户价值的提升呈现专业服务，为人居环境的改善描绘美好蓝图。

公司地址：上海市浦东新区东方路989号中达广场12楼
邮政编码：200122
公司总机：021-68758810
经营热线：021-68755817
公司传真：021-68758813
电子邮箱：csaa@shzjy.com
公司网址：www.shzjy.com

长沙世茂广场

上海张江集电港B区3-6研发总部

上海张江集电港B区3-6研发总部

中国商飞研发中心

四川都江堰体育馆

www.bacdesign.com.cn

住宅景觀・私家庭院
Residential Landscape・Private garden

酒店景觀・城市公共景觀
Hotel Landscape・Urban Planning Design

旅游度假項目規劃
Resorts and Leisure Planning

建築景觀模型
Architecture model

長期誠聘景觀設計人才,誠邀專業人士加盟.

地址:廣州市天河區龍怡路1號農機物資公司綜合樓東梯六樓(郵編:510635)
Add:6th Floor, Building of Agri-Materials Company, No.91, Longyi Rd., Tianhe Dist., Guangzhou(P. C. 510635)
電話Tel: 020-87569202 (0)13688860979/13360588555 Email:bacdesign@126.com

广州邦景园林绿化设计有限公司
GUANGZHOU BONJING LANDSCAPE DESIGN CO.,LTD

经营范围
城市景观规划 / 居住环境 / 企业环境 / 公园旅游区 / 园林工程

设计理念
潜心研究 / 用心规划 / 细心设计 / 精心实施

企业宗旨
设计创造品牌 / 服务雕琢精品 / 合作实现共赢

技艺——秉承着"国际化的设计视野，本土化的技术服务"理念。创造性地解决问题是邦景工作内容的核心，不管项目的规模有多大或项目的地点在哪里，邦景都会努力的找出问题的解决方案，以便赋予项目最为和谐的空间品质。

品质——秉承以质量求生存，以品质求发展的经营理念，以快捷的反应、严格的道德标准、精湛的技术水平、合理而具竞争力的价格和完善的售后服务，来适应未来市场发展的需要。

高效——及时高效的把控项目的整体进展。有序、明晰、详尽、高效的施工图编制，是邦景协助客户全面掌控景观工程的有效保障。

服务——对人文的关怀，对设计的热情，对业主的尊敬，对品位的追求，成就了邦景的创意无限和精诚服务，以实现对业主最为便捷与最有效的设计服务。

经验——丰富精湛的实战经验能确保项目的顺利进行，汇远见之睿智，扬开拓之激情，创天人合一的完美景致。

过去的荣誉，我们一起分享；未来的辉煌，我们将一起创造。
我们开放的平台一定让您无穷的创造力得以尽情施展。
随着业务的快速发展，现诚邀精英加盟发展。
地　址：广州天河北路175号祥龙花园晖祥阁2504/05
电话/传真：+86-20-87510037/38468069
邮箱：bonjing123@163.com
BOLG：@邦景园林　　QQ：120879245

創源國際
設計顧問有限公司
ORIGIN DESIGN & CONSULTING INTERNATIONAL CO., LIMITED

上海金创源建筑设计事务所有限公司

创源国际设计顾问有限公司（ORIGIN DESIGN & CONSULTING INTERNATIONAL CO,LIMITED）是一间注册在英国，面向国际的规划建筑设计机构.公司2004年进入中国大陆设计市场，合资成立上海金创源建筑设计事务所有限公司，并获得建设部授予的建筑设计事务所（有限公司形式）甲级资质。公司专注于高端住区及商业综合体设计领域，凭借先进的设计理念和全过程的服务管理，取得了骄人业绩。

ORIGIN DESIGN & CONSULTING INTERNATIONAL CO., LIMITED is a registered in the UK and international regulations Planning the architectural design firms. 2004 to enter the Chinese mainland design market, a joint venture in Shanghai Chong Yuen Construction Design Office Co., Ltd., and granted by the Ministry of Construction, Architectural Design Services (Limited companies) Grade A qualification. Company focused on high-end neighborhoods and commercial complex located Meter area, with advanced design concepts and the whole process of service management, and achieved remarkable results.

地址：上海杨浦区黄兴路1858号东上海中心701-703室
邮编：200433
电话：021-55062106,021-55092366
传真：021-55062106-807
网址：www.odci.com.cn
邮箱：workbox2001@odci.com.cn

专业·创新·共赢
www.odci.com.cn

| 住宅地産 | 商業地産 | 旅游地産 | 度假酒店 | 市政規劃 | 校園規劃 | 公園規劃 | 區域規劃 |

法國建築師協會會員單位

美國靈頓建築景觀設計有限公司是專業從事城市規劃和城市設計、風景區與公園景觀規劃、風景旅游渡假景觀、酒店環境景觀、高尚住宅小區景觀及工程設計的國際知名公司。深圳靈頓建築景觀設計有限公司是其專為中國地區設置的設計顧問公司，公司以國外多位設計師為公司主幹，美國公司作為主要規劃整體和宏觀控制，以共同協作，共同致力于同一項目上，使設計工程在整體規劃及局部處理上都得到精心設計。公司以全新的理念引導市場，以專業的服務介入市場，以全程的服務方式開拓市場，使公司得到穩步和迅速的發展，業務範圍不斷擴大。尤其在高尚住宅小區景觀設計中取的優异的成績，其設計工程多次獲國家和地區的獎勵。

http://www.szld2005.com   中國 深圳 福田區 紅荔路花卉世界313號

TEL：(0755) 8621 0770   FAX：(0755) 8621 0772   P.C：518000   EMAIL：szld2000@163.com   179049195@QQ.com

美國靈頓建築景觀設計有限公司   深圳靈頓建築景觀設計有限公司   雲南靈頓園林綠化工程有限公司

奥德景观
LUCAS DESIGN GROUP

西安阳光城-西西安小镇

深圳市奥德景观规划设计有限公司

公司坐落于著名的蛇口湾畔，深圳最有影响力的创意设计基地：南海意库；
公司前身为深圳市卢卡斯景观设计有限公司，是由 2003 年成立于香港的卢卡斯联盟（香港）国际设计有限公司在世界设计之都：深圳设立的中国境内唯一公司。
于 2012 年 1 月获得中华人民共和国国家旅游局正式认定：旅游规划设计乙级设计资质。

居住区景观与规划设计（含旅游地产）
商业综合体景观与规划设计（含购物公园、写字楼及创意园区）
城市规划及空间设计
市民公园设计
酒店与渡假村景观规划与设计
旅游策划及规划设计

倚重当下的中国的文化渊源结合世界潮流，尊重地域情感，在中国打造具强烈地域特征的、风格化的、国际化的，具前瞻性、可再生的的城市景观、人居环境、风情渡假区及自然保护区。

地　址：深圳市南山区蛇口兴华路南海意库 2 栋 410 室
电　话：0755-86270761
传　真：0755-86270762
邮　箱：lucasgroup_lucas@163.com
网　址：www.lucas-designgroup.com

前言 EDITOR'S NOTE

集中高效 品质至上
Concentrated, High Efficient and Quality First

商业综合体集城市中的商业、办公、居住、酒店、文化、休憩、交通等于一体，创造出一种多功能、复合型、高效率的建筑综合体。商业综合体是商业地产的高端发展模式，它涵盖了城市发展所需的各种要素和功能。当下，商业综合体的发展模式也成为我国商业地产发展中最为热门的模式，甚至有人形容："凡是有城市的地方就有商业综合体"。

从设计的角度来说，商业综合体较之单体建筑的设计更为复杂，并且要求更高的品质。设计者不仅要从建筑的角度思考，还需兼顾商业目的。同时，其定位、功能设置等还应从城市的角度出发，考虑到城市的总体规划。设计中还需关注到商业综合体与周边城市交通体系的关系，要建立与城市空间相协调的交通纽带，营造宜人的城市环境。此外，各类高科技、高智能设施的应用，可使商业综合体的设计更为高效、安全、人性化。

如今，商业综合体已然成为城市生活的重要角色，对城市的发展产生了巨大的影响。本期专题特别精选多个优秀商业综合体案例，与您共同探讨商业综合体设计中的方方面面。

Commercial complex gathers commerce, working, living, hotel, culture, recreation and transportation in one, creating a building complex with multi functions, compound type and high efficiency. Commercial complex is a high-end development pattern for commercial real estate development, which covers all kinds of all kinds of elements and functions functions needed in urban development. Now, the development pattern of commercial complex has become the most popular development pattern in Chinese commercial real estate, even someone said that "where is a city, where is with commercial complex".

From the perspective of design, compared with single building, commercial complex is more complicated and it requires higher standard quality. Designers not only think about construction but also consider commercial objectives. In the meanwhile, its positioning and function setting etc. should orient from the city overall planning. And the relationship between commercial complex and surrounding urban traffic system needs to be under consideration, so as to build traffic links coordinating with urban space and to create a attractive urban environment. Apart from that, the use of different kinds of high technologies and high intelligent facilities would make commercial complex design more effective, more security and more human-friendly.

Nowadays commercial complex plays an important role in city life, resulting in huge impact on city development. Therefore, in this issue multiple excellent projects of commercial complex are chosen to discuss with you about aspects of commercial complex design.

jiatu@foxmail.com

NEWHOUSE 图解地产与设计

2012年　总第43期

面向全国上万家地产商决策层、设计院、建筑商、材料商、专业服务商的精准发行

指导单位 INSTRUCTION UNIT
亚太地产研究中心

出品人 PUBLISHER
杨小燕 YANG XIAOYAN

主编 CHIEF EDITOR
王志 WANG ZHI

副主编 ASSOCIATE EDITOR
熊冕 XIONG MIAN

编辑记者 EDITOR REPOTERS
唐秋琳 TANG QIULIN
钟梅英 ZHONG MEIYING
胡明俊 HU MINGJUN
康小平 KANG XIAOPING
吴辉 WU HUI
曾伊莎 ZENG YISHA
曹丹莉 CAO DANLI
朱秋敏 ZHU QIUMIN
王盼青 WANG PANQING

设计总监 ART DIRECTORS
杨先周 YANG XIANZHOU
何其梅 HE QIMEI

美术编辑 ART EDITOR
詹婷婷 ZHAN TINGTING

国内推广 DOMESTIC PROMOTION
广州佳图文化传播有限公司

市场总监 MARKET MANAGER
周中— ZHOU ZHONGYI

市场部 MARKETING DEPARTMENT
方立平 FANG LIPING
熊光 XIONG GUANG
王迎 WANG YING
杨先凤 YANG XIANFENG
熊灿 XIONG CAN
刘佳 LIU JIA

图书在版编目（CIP）数据
新楼盘. 商业综合体：汉英对照 / 佳图文化主编.
——北京：中国林业出版社, 2012.11
ISBN 978-7-5038-6822-1

Ⅰ.①新... Ⅱ.①佳... Ⅲ.①建筑设计-中国-现代-图集 Ⅳ.①TU206

中国版本图书馆CIP数据核字(2012)第020869号
出版：中国林业出版社
主编：佳图文化
责任编辑：李顺 许琳
印刷：利丰雅高印刷(深圳)有限公司

特邀顾问专家 SPECIAL EXPERTS (排名不分先后)

赵红红 ZHAO HONGHONG	范勇 FAN YONG
王向荣 WANG XIANGRONG	赵士超 ZHAO SHICHAO
陈世民 CHEN SHIMIN	孙虎 SUN HU
陈跃中 CHEN YUEZHONG	梅卫平 MEI WEIPING
邓明 DENG MING	林世彤 LIN SHITONG
冼剑雄 XIAN JIANXIONG	熊冕 XIONG MIAN
陈宏良 CHEN HONGLIANG	周原 ZHOU YUAN
胡海波 HU HAIBO	李焯忠 LI ZHUOZHONG
程大鹏 CHENG DAPENG	原帅让 YUAN SHUAIRANG
范强 FAN QIANG	王颖 WANG YING
白祖华 BAI ZUHUA	周敏 ZHOU MIN
杨承刚 YANG CHENGGANG	王志强 WANG ZHIQIANG / DAVID BEDJAI
黄宇奘 HUANG YUZANG	陈英梅 CHEN YINGMEI
梅坚 MEI JIAN	吴应忠 WU YINGZHONG
陈亮 CHEN LIANG	曾繁柏 ZENG FANBO
张朴 ZHANG PU	朱黎青 ZHU LIQING
盛宇宏 SHENG YUHONG	曹一勇 CAO YIYONG
范文峰 FAN WENFENG	冀峰 JI FENG
彭涛 PENG TAO	滕赛岚 TENG SAILAN
徐农思 XU NONGSI	王毅 WANG YI
田兵 TIAN BING	陆强 LU QIANG
曾卫东 ZENG WEIDONG	徐峰 XU FENG
马素明 MA SUMING	张奕和 EDWARD Y. ZHANG
仇益国 QIU YIGUO	郑竞晖 ZHENG JINGHUI
李宝章 LI BAOZHANG	刘海东 LIU HAIDONG
李方悦 LI FANGYUE	凌敏 LING MIN
林毅 LIN YI	谢锐何 XIE RUIHE
陈航 CHEN HANG	

编辑部地址： 广州市海珠区新港西路3号银华大厦4楼
电话： 020—89090386/42/49、28905912
传真： 020—89091650

北京办： 王府井大街277号好友写字楼2416
电话： 010—65266908　　**传真：** 010—65266908

深圳办： 深圳市福田区彩田路彩福大厦B座23F
电话： 0755—83592526　　**传真：** 0755—83592536

协办单位 CO—ORGANIZER

广州市金冕建筑设计有限公司　熊冕 总设计师
地址：广州市天河区珠江西路5号国际金融中心主塔21楼06—08单元
TEL：020—88832190　88832191
http://www.kingmade.com

AECF 上海颐朗建筑设计咨询有限公司　巴学天 上海区总经理
地址：上海市杨浦区大连路970号1308室
TEL：021—65909515　　FAX：021—65909526
http://www.yl-aecf.com

WEBSITE COOPERATION MEDIA
网站合作媒体

 搜房网

副理事长单位 DEPUTY CHAIRMAN

华森建筑与工程设计顾问有限公司　邓明　广州公司总经理
地址：深圳市南山区滨海之窗办公楼6层
　　　广州市越秀区德政北路538号达信大厦26楼
TEL：0755—86126888　020—83276688
http://www.huasen.com.cn　E—mail:hsgzaa@21cn.net

广州市瀚华建筑设计有限公司　冼剑雄　董事长
地址：广州市天河区黄埔大道中311号羊城创意产业园2—21栋
TEL：020—38031268　FAX：020—38031269
http://www.hanhua.cn
E—mail：hanhua—design@21cn.net

上海中建建筑设计院有限公司　徐峰　董事长
地址：上海市浦东新区东方路989号中达广场12楼
TEL：021—68758810　FAX：021—68758813
http://www.shzjy.com
E—mail：csaa@shzjy.com

常务理事单位 EXECUTIVE DIRECTOR OF UNIT

深圳市华域普风设计有限公司　梅坚　执行董事
地址：深圳市南山区海德三道海岸城东座1306—1310
TEL：0755—86290985　FAX：0755—86290409
http://www.pofart.com

华通设计顾问工程有限公司
地址：北京市西城区西直门南小街135号西派国际C—Park3号楼
TEL：8610—83957395　FAX：8610—83957390
http://www.wdce.com.cn

天萌（中国）建筑设计机构　陈宏良　总建筑师
地址：广州市天河区员村四横路128号红专厂F9栋天萌建筑馆
TEL：020—37857429　FAX：020—37857590
http://www.teamer—arch.com

GVL国际怡境景观设计有限公司　彭涛　中国区董事及设计总监
地址：广州市珠江新城华夏路49号津滨腾越大厦南塔8楼
TEL：020—87690558　FAX：020—87697706
http://www.greenview.com.cn

天友建筑设计股份有限公司　马素明　总建筑师
地址：北京市海淀区西四环北路158号慧科大厦7F（方案中心）
TEL：010—88592005　FAX：010—88229435
http://www.tenio.com

R—LAND 北京源树景观规划设计事务所　白祖华　所长
地址：北京朝阳区朝外大街怡景园5—9B
TEL：010—85626992/3　FAX：010—85625520
http://www.ys—chn.com

奥雅设计集团　李宝章　首席设计师
深圳总部地址：深圳蛇口南海意库5栋302
TEL：0755—26826690　FAX：0755—26826694
http://www.aoya—hk.com

北京袁亚国际建筑设计有限公司　赵士超　董事长
地址：北京市朝阳区琨莎中心1号楼1701
TEL：010—65797775　FAX：010—84682075
http://www.hygjjz.com

广州山水比德景观设计有限公司　孙虎　董事总经理兼首席设计师
地址：广州市天河区珠江新城临江大道685号红专厂F19
TEL：020—37039822/823/825　FAX：020—37039770
http://www.gz—spi.com

奥森国际景观规划设计有限公司　李焯忠　董事长
地址：深圳市南山区南海大道粤海路动漫园7栋5楼
TEL：0755—26828246 86275795　FAX：0755—26822543
http://www.oc—la.com

广州市四季园林设计工程有限公司　原帅让　总经理兼设计总监
地址：广州市天河区龙怡路117号银汇大厦2505
TEL：020—38273170　FAX：020—86682658
http://www.gz—siji.com

深圳市雅蓝图景观工程设计有限公司　周敏　设计董事
地址：深圳市南山区南海大道2009号新能源大厦A座6D
TEL：0755—26650631/26650632　FAX：0755—26650623
http://www.yalantu.com

深圳市佰邦建筑设计顾问有限公司　迟春儒　总经理
地址：深圳市南山区兴工路8号美年广场1栋804
TEL：0755—86229594　FAX：0755—86229559
http://www.pba—arch.com

北京新纪元建筑工程设计有限公司　曾繁柏　董事长
地址：北京市海淀区小马厂6号华天大厦20层
TEL：010—63483388　FAX：010—63265003
http://www.bjxinjiyuan.com

北京博地澜屋建筑规划设计有限公司　曹一勇　总设计师
地址：北京市海淀区中关村南大街31号神舟大厦8层
TEL：010—68118690　FAX：010—68118691
http://www.buildinglife.com.cn

HPA上海海波建筑设计事务所　陈立波、吴海青　公司合伙人
地址：上海中山西路1279弄6号楼国峰科技大厦11层
TEL：021—51168290　FAX：021—51168240
http://www.hpa.cn

香港华艺设计顾问（深圳）有限公司　林毅　总建筑师
地址：深圳市福田区华富路航都大厦14、15楼
TEL：0755—83790262　FAX：0755—83790289
http://www.huayidesign.com

哲思（广州）建筑设计咨询有限公司　郑竞晖　总经理
地址：广州市天河区天河北路626号保利中宇广场A栋1001
TEL：020—38823593　FAX：020—38823598
http://www.zenx.com.au

理事单位 COUNCIL MEMBERS（排名不分先后）

广州市柏澳景观设计有限公司　徐农思　总经理
地址：广州市天河区广园东路2191号时代新世界中心南塔2704室
TEL：020—87569202　FAX：020—87635002
http://www.bacdesign.com.cn

中房集团建筑设计有限公司　范强　总经理/总建筑师
地址：北京市海淀区百万庄建设部院内
TEL：010—68347818

北京奥思得建筑设计有限公司　杨承冈　董事总经理
地址：北京朝阳区东三环中路39号建外SOHO16号楼2903~2905
TEL：86—10—58692509/19/39　FAX：86—10—58692523

陈世民建筑师事务所有限公司　陈世民　董事长
地址：深圳市福田中心区益田路4068号卓越时代广场4楼
TEL：0755—88262516/429

广州嘉柯园林景观设计有限公司　陈航　执行董事
地址：广州市珠江新城华夏路49号津滨腾越大厦北塔506—507座
TEL：020—38032521/23　FAX：020—38032679
http://www.jacc—hk.com

侨恩国际（美国）建筑设计咨询有限公司
地址：重庆市渝北区龙湖MOCO4栋20—5
TEL：023—88197325　FAX：023—88197323
http://www.jnc—china.com

CDG国际设计机构　林世彤　董事长
地址：北京海淀区长春路11号万柳亿城中心A座10/13层
TEL：010—58815603 58815633　FAX：010—58815637
http://www.cdgcanada.com

广州市圆美环境艺术有限公司　陈英梅　设计总监
地址：广州市海珠区宝岗大道杏坛大街56号二层之五
TEL：020—34267226 83628481　FAX：020—34267226
http://www.gzyuanmei.com

上海唯美景观设计工程有限公司　朱黎青　董事、总经理
地址：上海市徐虹中路20号2—202室
TEL：021—61122209　FAX：021—61139033
http://www.wemechina.com

上海金创源建筑设计事务所有限公司　王毅　总建筑师
地址：上海市杨浦区黄兴路1858号701—703室
TEL：021—55062106　FAX：021—55062106—807
http://www.odci.com.cn

深圳灵顿建筑景观设计有限公司　刘海东　董事长
地址：深圳福田区红荔路花卉世界313号
TEL：0755—86210770　FAX：0755—86210772
http://www.szld2005.com

深圳市奥德景观规划设计有限公司　凌敏　董事总经理、首席设计师
地址：深圳市南山区蛇口海上世界南海意库2栋410#
TEL：0755—86270761　FAX：0755—86270762
http://www.lucas—designgroup.com

广州邦景园林绿化设计有限公司　谢锐何　董事及设计总监
地址：广州市天河北路175号祥龙花园晖祥阁2504/05
TEL：020—87510037　FAX：020—38468069
http://www.bonjinglandscape.com

目录 CONTENTS

011　前言 EDITOR'S NOTE

016　资讯 INFORMATION

名家名盘 MASTER AND MASTERPIECE

020　北京金科帕提欧一期：极具西班牙风情的诗意化人文住区
　　　ROMANTIC CULTURAL COMMUNITY IN SPANISH STYLE

028　上海中海瀛台二期：端庄、高雅的新古典主义江景社区
　　　ELEGANT AND EXQUISITE NEOCLASSIC WATERFRONT COMMUNITY

专访 INTERVIEW

034　商业地产要整合好策划、营销与设计之间的关系
　　　——访深圳市佰邦建筑设计顾问有限公司董事、总建筑师　王志强
　　　COMMERCIAL PROPERTY SHOULD INTEGRATE THE RELATIONSHIP AMONG SCHEME, MARKETING AND DESIGN WELL

038　"我们憧憬的未来"——诗意栖居
　　　——访广州山水比德景观设计有限公司执行董事、总设计师　孙虎
　　　THE FUTURE WE LOOK FORWARD TO — THE POETIC INHABITATION

新景观 NEW LANDSCAPE

042　旧金山理查森公寓：生态节能、绿色高效的住区景观
　　　ECO RESIDENTIAL LANDSCAPE

046　广州汇东国际花园：生态、人文气息的高品位心灵居所
　　　ECOLOGICAL RESIDENCE WITH HUMANISM AND HIGH QUALITY

050　广州大一山庄：水月云天　诗意人间
　　　MOON WATER CLOUD, POETRY WORLD

054　长沙万科金域华府（一、二期）：静谧　简约　高贵　典雅
　　　TRANQUIL SIMPLE NOBLE ELEGANT

020

042

专题 FEATURE

- 062 浅谈混合用途建筑的历史与未来
  MIXED-USE DEVELOPMENT: PAST, PRESENT AND FUTURE

- 066 关于商业综合体的设计关注点
  CONCERNS ON THE DESIGN OF COMMERCIAL COMPLEX

- 068 浅谈商业综合体的设计及发展
  WITH CONCISE REMARKS ON THE DESIGN AND DEVELOPMENT OF COMMERCIAL COMPLEX

- 074 巴林城市中心：造型独特、蕴含海滨元素的顶级商业空间
  TOP COMMERCIAL SPACE WITH SPECIAL FORM AND SEASIDE ELEMENTS

- 080 Kameleon商住综合体：水平切割的表皮 围合形式的布局
  HORIZONTALLY DISSECTED SKIN AND ENCLOSED LAYOUT

- 090 贵州遵义唯一国际：开放立体、依山而建的景区化城市综合体
  OPEN AND SCENIC URBAN COMPLEX BUILT NEARBY THE MOUNTAIN

- 096 长沙世茂广场：有机组合的建筑形态 简约动感的城市空间
  ORGANICALLY COMBINED ARCHITECTURAL FORM CONTRACTED AND DYNAMIC URBAN SPACE

- 100 张家港休闲购物公园：充满秩序与逻辑感的多变城市综合体
  CHANGEABLE URBAN COMPLEX FULL OF ORDERLINESS AND LOGICAL SENSE

新特色 NEW CHARACTERISTICS

- 108 天津卡梅尔：精致内敛的西班牙南加州风格住区
  DELICATE AND RESTRAINED RESIDENCE IN SPANISH SOUTHERN CALIFORNIA STYLE

新空间 NEW SPACE

- 116 ICON·大源国际中心售楼部：缔造"德系精工"品质现代空间
  BUILDING A MODERN SPACE WITH "GERMAN ELABORATION"

新创意 NEW IDEA

- 120 海港公寓大楼：独特"马赛克"图案立面的滨水居所
  WATERFRONT RESIDENCE WITH DISTINCTIVE MOSAIC FACADE

商业地产
COMMERCIAL BUILDINGS

- 130 赛特71商住公寓：历史与现代元素的碰撞融合
  THE INTEGRATION OF PAST AND MODERN

- 136 云龙数码科技城规划设计：绿色生态 科技新城
  GREEN AND ECOLOGICAL, NEW TECHNOLOGY CITY

- 142 CIB生物医学研究中心：仿生学原理在建筑中的综合运用
  IMPLEMENTATION OF BIOMIMICRY IN ARCHITECTURE

INFORMATION | 资讯/地产

财政部追加2012年公租房棚改专项资金50亿

财政部网站10月8日公布，中央财政追加下达2012年中央补助公共租赁住房和城市棚户区改造专项资金50亿元人民币，用于公共租赁住房和城市棚户区改造相关配套基础设施建设支出。截止目前，中央财政已累计下达2012年公共租赁住房和城市棚户区改造补助资金987亿元。

CHINA MINISTRY OF FINANCE PROVIDES ADDITIONAL SPECIAL FUNDS REACHING TO 5 BILLION YUAN FOR PUBLIC RENTAL HOUSING AND TRANSFORMATION OF URBAN SHANTYTOWNS

On Oct.8th, the Ministry of Finance website published that China Ministry of Finance provides special additional funds reaching to 5 billion yuan, which is used for the cost of public rental housing and the spending of relevant infrastructure for the transformation of urban shantytowns. So far, China Ministry of Finance has already distributed 98.7billion for 2012 public rental housing and transformation of urban shantytowns.

深圳市民购房可申请住房公积金贷款

9月25日，深圳市住房公积金管理中心发布公告称，深圳市住房公积金贷款业务将于2012年9月28日上线试运行。此外，深圳规定五年以下(含五年)的住房公积金贷款年利率为4%；五年以上的年利率为4.5%，比目前商业银行个人住房贷款基准利率低了2.05个百分点。

SHENZHEN CITIZENS ARE ELIGIBLE FOR HOUSING PROVIDENT FUND WHILE BUYING HOUSES

Shenzhen Housing Accumulation Fund Administration Centre has issued on Sep.25th that the trial operation of Shenzhen housing provident fund would start on Sep.28th. In addition, there are rules in Shenzhen that annual interest of housing provident fund under 5 years (included) is 4%, and annual interest of housing provident fund over 5 years is 4.5% which is 2.05 percentage point below current commercial banks' benchmark interest rate of personal housing loan.

税务总局：房屋交换价格相等可免交契税

据国家税务总局纳税服务司介绍，土地使用权交换、房屋交换，交换价格不相等的，由多交付货币、实物、无形资产或者其他经济利益的一方缴纳税款。交换价格相等的，免征契税。

SAT OF CHINA: THE EXCHANGED HOUSES OF EQUAL VALUE MAY BE EXEMPT FROM DEED TAX

According to China's State Administration of Taxation, if the exchanged land use-right and the swapped houses are not equal value, the one who deliver more currency, in kind, intangible assets or other economic interests will pay taxes. But the deed tax will be spared as long as the exchanged items are of equal value.

贵阳楼市新政：首次购房后可享本市户籍待遇

贵阳市人民政府日前发布《关于推进保障性安居工程和棚户区城中村改造及重点项目建设的若干措施(试行)》。《措施》规定，凡在贵阳市购买商业、办公用房和首次购买住房的，根据本人意愿可持房屋所有权证或备案的购房合同在购房区域办理入户手续或居住证，享受本市户籍人口就业、入学和就医等同等待遇，不受入住时间和购房面积的限制。

GUIYANG NEW PROPERTY POLICY: FIRST-TIME HOME BUYER MAY ENJOY SAME TREATMENTS AS THE PEOPLE WITH THE HOUSEHOLD REGISTRATION IN THIS CITY

Guiyang People's Government issued <<Measures about promoting government-subsidized housing project and key project construction and transformation in urban village of shantytowns (for trial)>> a few days ago. The <<Measures>> set out to buyers, who purchase commercial premises, office properties and buy housing for the first time, and, on voluntary basis, summits identification document or residence permits to the region government office with building droit card or a registered sales contract, may enjoy same treatments as the people with the household registration in this city in the aspects such as employment, child's education and medical care without limitation of move in date and house purchase area .

9月房地产百城价格指数涨幅收窄至0.17%

10月8日，中国指数研究院发布的数据显示，9月全国100个城市（新建）住宅平均价格为8 753元/平方米，环比8月上涨0.17%。这是自今年6月止跌后连续第4个月环比上涨，不过涨幅继续收窄。

THE 100-CITIES INDEX OF REAL ESTATE PARED GAINS TO 0.17% IN SEP

On Oct.8th, figures from China Index Academy showed that in September the average price of new housings in 100 cities throughout the country was RMB8, 753 per square meter, increased 0.17% compared with August. This was an increase for 4 consecutive months since June this year, but the gains continue to subside.

万科佳兆业佛山再争地

10月8日，佛山市南海区就迎来了十月的第一场土地拍卖会。当日，佛山市南海区桂城街道共有四块土地进行拍卖。据统计，当日出让的四块土地总出让面积为159 602.1m²，总起拍价共计167 710万元。最终，该四块地分别由万科和佳兆业竞得，总成交价达24.45亿元。

VANKE AND KAISA BID FOR LAND PURCHASES AGAIN

On Oct.8th, the first land auction of this month was held in Nanhai District, Foshan City. At that day, there are 4 plots of land being listed for auction. According to statistics, the total transferred land area is 159,602.1m², and the starting price amounts to 1.6771 billion. Finally, those four plots of land are purchased by Vanke and Kaisa separately, and a total value of 2.445 billion.

深圳秋交会纪实：楼市低迷

10月5日，深圳秋交会落下帷幕。据深圳市规土委网站统计，在为期5天的展会中，深圳一手住宅成交仅34套，相比去年秋交会成交面积跌幅更是超过五成。秋交会成交的低迷，导致深圳楼市"银十"开局不利，更为接下来的深圳楼市增添了更多不确定因素。

ON-THE-SPOT REPORT OF SHENZHEN AUTUMN REAL ESTATE TRADE FAIR: HOUSING MARKET SLUMPS

2012 Shenzhen Autumn Real Estate Trade Fair ends on Oct.5th.Statistics from Urban Planning Land and Resources Commission of Shenzhen Municipality indicates that only 34 new homes have been sold in the five-day fair, the deal area of which suffers declines of more than 50 percent compared to last autumn fair. The market downturn in the fair results in a weak start for the Shenzhen property market in this silver October; what is more, adding more uncertainty to the next development of Shenzhen housing market.

54城签约套数环比剧降七成

"十一"黄金周期间，各地楼市成交量均有不同程度下调。据统计，黄金周期间全国54城市住宅签约量环比大降7成，签约套数合计仅为1.6万套。对此，住建部政策研究中心副主任王珏林表示，房地产调控绝不放松，年内全国楼市将以平稳为主。

THE QUANTITY OF SIGNED HOUSING CONTRACT IN 54 CITIES HAS DRAMATICALLY DECLINED 70%

Over the recent seven-day National Day Holiday, sales volumes have dropped to verifying degrees across the country. According to statistics, during this holiday the quantity of signed housing contract, which is only 16,000, has dramatically declined 70% on a sequential basis in 54 cities throughout the country. To this, Mr. WANG Juelin, deputy head of MHURC's research arm, expressed that there will not be even the slightest faltering in the property-market curbs and the property market will be dominated by smooth in this year.

专家称房产暴利时代已终结

今年的"十一"黄金周期间，房地产市场成交量并不理想，各地楼市陷入"休假"状态，"金九银十"旺季开局并不明显。北京理工大学房地产研究所所长周毕文表示，房地产暴利时代已经终结，未来10月份楼市将维持量价齐稳的状态。

EXPERT CLAIMED HOUSING OF PROFITEERING ERA ENDS

Over the National Day Holiday in this year, transaction volumes of housing market are not ideal. And the property market at all area slumbers. As a result, this peak season in "gold September silver October" started not very well. Bi-wen Zhou, director of BIT Real Estate Institute, said housing of profiteering era already ended, and price and volumes of housing market would both remain stable throughout this October.

INFORMATION 资讯/设计

发光住宅

这个住宅位于一个斜坡前方，稍微嵌进后方的山体，楼梯装饰有带有金属光泽的铝制肋条钢丝网。建筑最上方两层每层的拐角处都有一个窗户，住户从室内能以良好的角度欣赏周围的景色。整个建筑充分利用可再生能源，包括地热泵、室内通风系统和热量交换机等。

Reflecting Cube

On a recessed natural stone base abuts a cube made of shining aluminium rib mesh. Four corner glazings were cut into this cube. Two of them create a visual connection to the middle-age ruin Windeck. The entry, an office and the garage are located in the semi-basement. The living area in the top level opens up to the terrace and the edge of the forest with a big window. The topic of renewable energy resources is becoming more and more important. Because of this, the architect decided to use a geothermal heat pump and a controlled living room ventilation with ground heat exchangers.

多变式公寓

公寓位于一片吸引人的区域，是一片开放的私人庭院周围建筑群的一部分，私人庭院地面较高于周围。公寓楼最大的特色就是大量凸窗和阳台，赋予整栋建筑宏伟的，雕塑般的气质。大楼外表由黄色水泥契合的砖块构成。

Yoo Apartments

The residential building is part of an ensemble around an open, raised private courtyard. The urban planning framework was given. The architectural characteristics present large oriels and balconies as a generous, sculptural figure with fine apartments. The facade consists of faced brickwork as a hanging facade structure conceived with limed bricks in beige.

K住宅

从外表看，这栋住宅就是一个9m高的密闭混凝土结构，北面临街，两面表面几乎是不透光的。整个建筑是用预制混凝土板组成的，再加上结实的木制格子结构，看似相互矛盾的材料，为房间增添了东方气息和温馨感。

House K

The design is a concrete block, 9 meters height without visible openings. The north elevation facing the street and both side facades seem completely opaque and yet they are not alienated to their environment. The entire structure is covered with a uniform system of prefabricated exposed concrete panels, which are integrated with heavy wood Latticework – A reminder to the traditional oriental element – the eastern trellis ("mashrabia"). The combination of materials and distribution arrangements add warmth, and ease the rigid system.

K.I.S.S.住宅楼

这栋住宅楼共有46间公寓，分为三种户型，内部装修与住宅大楼外部整体都具有设计大胆新颖的特征。每间公寓的风格都不同，有经典风格、工业风格、时尚风格等等，为住户提供多样的居住体验。

K.I.S.S.

K.I.S.S. is a boldly realized residential concept on Badenerstrasse in Zurich presenting 46 maisonnette apartments with striking interior designs. The tenant is given the choice between three exemplary apartment types offering individual experiences. This creates a whole new living concept enabling the choice between apartments based on identical layouts but with their own character, carrying names such as CLASSIC, INDUSTRIAL and FUNKY. K.I.S.S. occupies a striking corner plot and thereby creates an independent building sculpture.

山脊路别墅

简单的建筑理念，配合陡坡地形和一棵造型简洁的茶树，共同构成了这幢布局有限，但功能齐全，与周围环境完美结合的建筑，赋予其独特的美感。别墅内分为公共区域和私人区域。别墅周围的自然环境较为荒凉，设计师建造了高出地面很多的露台平台，以减少斜坡对建筑的影响。

Ridge Road Residence

A simple brief and programming requirements, teamed with a dramatic site characterized by a steep slope and a single tea tree, enabled the design to become an exploration into enclosing the basic rituals of domestic life within restrained building forms…the form of the building becomes driven by the clients desire to separate the public and private zones of the residence. In stark contrast to the surrounding houses, which attempt to cancel out the sloping topography by creating a podium level at which the outdoor areas sit exposed high above ground level, our design for this house adopts a gentler strategy, with the building form spilling down the slope to terminate in a series of terraced decks.

FF别墅

别墅主体造型简洁，选用轻质材料，从外面看，仿佛是漂浮在木制基座上的结构。内部装饰给人带来无拘无束的感觉。整个建筑平面呈T字形，入口走廊两边是花园，车库入口主材料是木材，庭院开阔，视野通透。

FF House

The formal concept aims to display simple and lightweight items that observed from the outside seems to float on a wooden base. From inside the bodies give the feeling of being free and unsupported. The whole floor plan is in T shape, side gardens are arranged along the entrance corridor. Wood is the main material in the garage entrance. Courtyard is open to the side garden visually through a floor to ceiling window.

跳台滑雪道建筑

挪威霍尔门科伦地区一直以来以举办举世瞩目的滑雪盛会而闻名世界,在体育界与温布尔顿和温布利体育馆齐名。但是由于自然条件限制,跳台滑雪山坡高度较低,因此,奥斯陆政府在这里新开了这个滑雪跳台。

Holmenkollen Ski Jump

Along with Wimbledon and the Wembley Arena, Holmenkollen Ski Jump the world's most visited sports facility. Nevertheless as a ski jump, it is one of the smallest hills in the World Cup tournament, and in September 2005, the International Ski Federation decided that the current hill does not meet the standards to award the city the 2011 FIS Nordic World Ski Championships. In December 2005 Norway's Directorate of Cultural Heritage approved the demolition of the ski jump and in April 2007 the Oslo municipality announced an open international competition for a new ski jump.

无限大楼

由Kohn Pedersen Fox Associates事务所设计的无限大楼位于圣保罗新兴经济区,距离法利亚利马不远,建筑造型独特,极具未来感,是一栋全新的高级办公楼。大楼总高120m,有18层,形似一艘满帆前进的航船。

Infinity Tower

International architecture firm Kohn Pedersen Fox Associates (KPF) is pleased to announce the completion of Infinity Tower. Located in São Paulo's new financial district, steps away from Faria Lima, this unique, next-generation building sets a new standard for Class-A office in the Brazil market and has already attracted world-class attention. The building's beauty is matched by its world-class functionality.

布尔格斯艺术学校

艺术学校大楼位于一片居民区里,底层几何形状与所处位置相符合。上层是主入口,大厅和会议室,空间呈直角分布。底层作为基座,是由黑色砖块组成的。上下两层都有相互重叠的部分。整个建筑内部空间以模块化的结构分布,布局简单。

Art School

The building, situated within a block of residential buildings, adopts the geometrical shape of its location. A lower section on one floor follows the shape of the property. Above this there is a section of the building containing the entrance, main foyer and conference room, with an orthogonal footprint. Its outward appearance expresses place, and new order. The ground floor acts as a base, built in black clinker brickwork. Both the two sections and their respective uses overlap. This helps to achieve the sense of a large open space. The conference room can be extended, when an event so requires, to include the entrance hall, while the entrance hall in turn blends into the conference room when it is not in use.

阿联酋珍珠酒店

珍珠酒店与酋长国宫殿酒店离的很近,紧邻沙滩散步道,是一家顶级商务酒店。酒店造型气派独特,各种设施应有尽有,包括停机坪、餐厅、健身房、散步道等,形成一处综合的设施。酒店总共有365间客房。酒店水平面呈半圆形,外表结构呈螺旋形以椭圆结构为中心向上延伸。

Emirates Pearl Hotel

An expressive form – presenting a unique appearance from every perspective – and a luxurious and modern design concept as well as a comprehensive assortment of amenities, ranging from a helicopter pad with direct elevator access, various restaurants and wellness enticements to a marina complex, create a most comprehensive experience. All in all, Emirates Pearl Hotel has 365 rooms – from double bed to luxury suite – that offer comfortable accommodation for every guest. Two semi-circular hotel wings spiral upwards around an elliptical circulation core.

布莱迪斯大学招生中心

招生中心位于校园入口处不远的位置,设计师新建了另一个石灰岩和玻璃雕塑。招生中心首层能同时容纳100人,设计师设计了三个分开的等候区,等候区内气氛温馨,光线充足,其中一个还配有壁炉。

Admissions Center, Brandeis University

Designers linked the admissions building closely to the campus center by creating another sculptural design with limestone and abundant glass. The ground floor public space needed to be able to accommodate 100 visitors at a time. To create a more personal and intimate feel, designers designed three separate waiting areas rather than herd people into one large area. The waiting areas are warm and light-filled; one has a fireplace.

卢浮宫伊斯兰艺术画廊

卢浮宫新建的伊斯兰艺术画廊近期对外开放,画廊外庭院地面上方覆盖有这样的波浪状金色平面结构。结构周围的建筑正面属于新古典主义风格,它的支撑结构是由棋盘格式的三角形玻璃组成,正反两面都覆盖了阳极氧化铝网,形成金色的表层效果。

Department of Islamic Arts at Musée du Louvre

An undulating golden plane blankets the new Islamic art galleries at the Musée du Louvre in Paris, which opened to the public this weekend. The new gallery wing is surrounded by the neoclassical facades of the museum's Cour Visconti courtyard and has two of its three floors submerged beneath the ground. Tessellated glass triangles create the self-supporting curves of the roof and are sandwiched between two sheets of anodized aluminium mesh to create a golden surface both inside and out.

ROMANTIC CULTURAL COMMUNITY IN SPANISH STYLE | Beijing Jinke Patio Phase One

极具西班牙风情的诗意化人文住区 —— 北京金科帕提欧一期

项目地点：中国北京市
建筑设计：筑博设计集团（北京）股份有限公司
用地面积：83 964 m²
总建筑面积：171 274.59 m²
容 积 率：1.6
绿 地 率：30.0%

Location: Beijing, China
Architectural Design: Zhubo Design Group (Beijing) Co., Ltd.
Land Area: 83,964 m²
Total Floor Area: 171,274.59 m²
Plot Ratio: 1.6
Greening Rate: 30.0%

项目概况

金科帕提欧项目座落于昌平区小汤山温泉资源核心地带，紧邻龙脉温泉度假中心、小汤山疗养院，属于不可多得的汤泉养生之地。整个区域内绿树成荫，温榆河穿流其中，形成了极佳的天然人居环境，交通便捷。

规划布局

金科帕提欧的整体规划设计灵感来自于西班牙城市塞维利亚，在地块中用一条贯穿整个地块的水系将地块巧妙地分成了别墅区和洋房区，满足不同人群的高品质居住需求。

结合景观、噪声与土地价值等诸多设计条件，设计师将地块分为东部的别墅区，西部的洋房区和沿街的公寓区。水系从位于西南角的会所后的温泉泳池始源，贯穿整个地块。在街区设计中，强调规划尺度适宜，与人的活动以及建筑风格相协调，重视人的存在感和体验感。

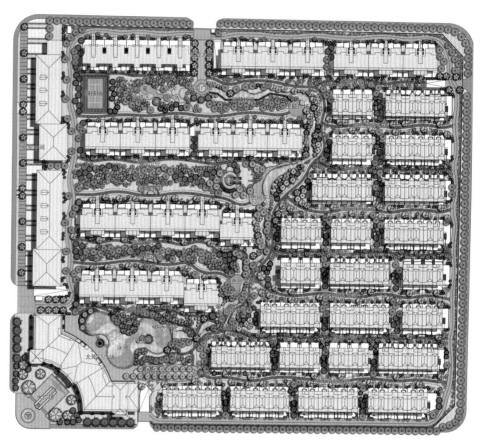

总平面图 Site Plan

MASTER AND MASTERPIECE | 名家名盘

建筑设计

项目的建筑风格源于西班牙庭院别墅，大量使用手工材料，呈现出原汁原味的西班牙风情。大量的露台和下沉庭院真正体现了西班牙风格的热情洋溢。露台作为室内空间的外延，让家最大程度地迎接阳光和空气。

"庭院"为别墅赋予了更多的内涵。由于阳光的经常光临，地下室也可以和窗外的庭院一样，展现出多姿多彩的居住灵感，同时温泉入户与园区温泉泳池更使得西班牙建筑美学与小汤山的自然天成完美融合。在细节上力求体现浓郁的西班牙风情，红色的坡屋顶，斑驳的文化石外墙与古典的拱门，复古工艺的铜质栏杆，庄严俊朗的阳光连廊，与静谧的池水、葱郁的园林融为一体，如同画卷。

户型设计

设计师打破了传统的别墅设计模式，充分利用竖向高度将功能空间划分成尺度更适宜的居住空间。会客区和次卧、主卧、儿女卧室区、工作室、客卧截然分开、用一部私家电梯联系起来，创造了高品位的居住空间。同时结合建筑造型，通过局部退台形成了阳光地下室和大露台，为创造富有情趣的生活提供了良好的居住环境。

景观设计

基于项目所在地块特定的城市区域位置，以及优越的自然环境条件，设计将风格化的建筑空间与自然环境因素有机结合起来，作为设计的出发点。因此提出了"打造诗意的人文居住模式，创造高品味的滨水生活"，并进行立体复合造景，使得四季景观画面层次分明，着力打造京城寓所罕有的公园级成品园林。

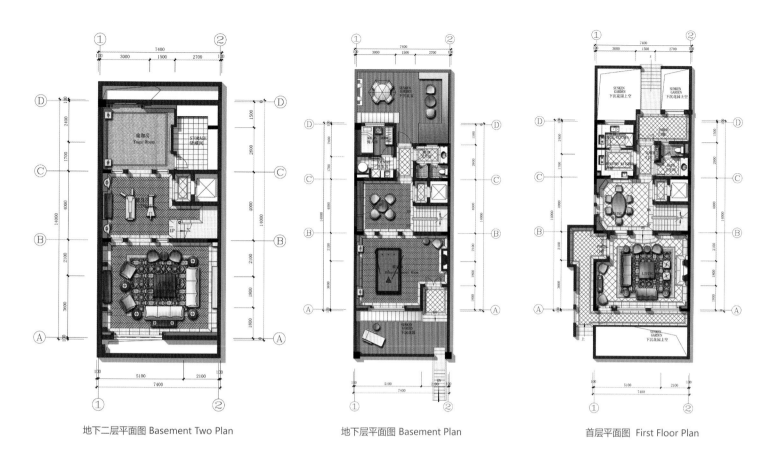

地下二层平面图 Basement Two Plan

地下层平面图 Basement Plan

首层平面图 First Floor Plan

NEW HOUSE _023

MASTER AND MASTERPIECE | 名家名盘

Profile
Project Jinke Patio is located in the central area of Xiaotangshan Hot Spring of Changping District, closely next to Longmai Hot Spring Vacation Center and Xiaotangshan Sanatorium, which is a precious hot spring land to preserve one's health. The whole place is covered by green shade with Wenyu River running through, which form a best natural residence and enjoys convenient transportation.

Planning and Layout
The inspiration of Jinke Patio came from the Spanish city Sevilla, which utilizes a water system to deliberately divide the entire area into villa area and foreign-style house area to meet different residential requirement of people.

After combining the conditions of landscape, noise and land value, architects decided to divide the east plot as villa area, west plot as

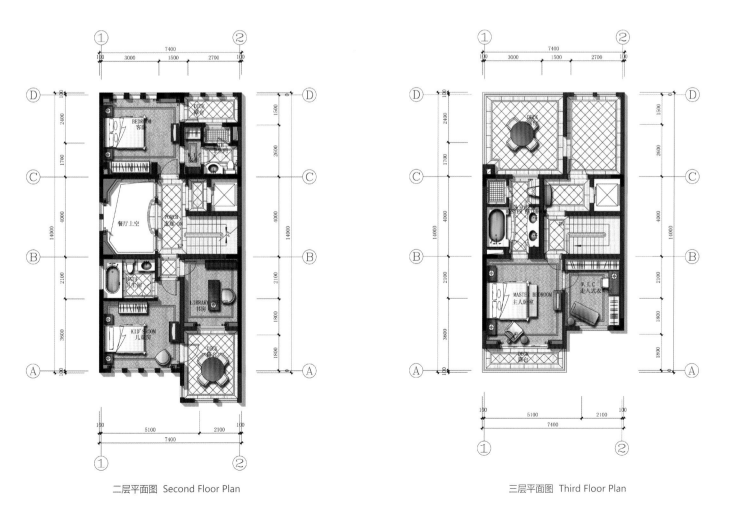

二层平面图 Second Floor Plan

三层平面图 Third Floor Plan

foreign-style house area and the apartment area along street. The water system starts from the hot spring pool in southwest corner and runs across the whole residential area. In design of city block, architects highlight the coordination of appropriate scale, human activities and architectural style and stress on the existence and experience of people.

Architectural Design

The architectural style of the project originated from Spanish courtyard villa, which applies a large amount of handmade materials to display authentic Spanish temperament. The design of terraces and submerged courtyards truly embody the passion and enthusiasm of Spanish style. As an extension of interior space, terrace provides the largest extent of air and sunshine to residents.

The courtyards have left villas with more connotations. With the frequent visiting of sunshine, the basement could become like courtyard and display varied and colorful residential inspiration. Meanwhile, the coming indoor of hot spring and the pool together provides a perfect combination for Spanish aesthetics and natural beauty of Xiaotangshan. The project tried to demonstrate heavy Spanish expressions through details like red slope-roof, mottled culture stone wall, classical arch door, copper rail imitating the ancient style, solemn sunny corridor, tranquil water and lush garden.

Type Design

Architects broke the traditional design style of villas but utilized vertical height to divide the space into residential space which provides more appropriate sizes for living. Visitor meeting area is completely separated from living rooms, working room and guest room and connected by a private elevator to create high quality living space.

Landscape Design

Based on the particular urban location of the project, the superior natural environment, architects organically combined the architectural space with natural elements as the starting point to promote the theme of "forging a poetic residence and creating high quality waterside life" and carry through three-dimensional complex landscape creation, which offers clear four-season views and rare garden-type residence among similar projects of Beijing.

ELEGANT AND EXQUISITE NEOCLASSIC WATERFRONT COMMUNITY | Shanghai Bay Line, Phase II

端庄、高雅的新古典主义江景社区 —— 上海中海瀛台二期

项目地点：中国上海市徐汇区
开 发 商：上海新海汇房产有限公司
建筑设计：天华建筑设计有限公司
总用地面积：83 241 m²
总建筑面积：130 871.16 m²
容 积 率：1.31

Location: Xuhui, Shanghai, China
Developer: Shanghai Xinhui Real Estate Co., Ltd.
Architectural Design: Tianhua Architecture
Total Land Area: 83,241 m²
Total Floor Area: 130,871.16 m²
Plot Ratio: 1.31

项目概况

建设基地位于上海市徐汇区，东面紧临黄浦江，西接龙吴路，北面隔规划市政道路龙瑞路与一期相邻。出基地沿龙吴路往南1 km即接A20公路与徐浦大桥。建筑功能主要为住宅小区，同时在用地内地块西侧分别规划一座10班幼儿园和30班小学以及菜场、商店等配套公建。

规划布局

项目在"高效、合理、出新"的原则下，延续一期规划脉络注重对基地自身环境特点的挖掘与开发，形成突出的特点。由于黄浦江景是本项目最有利的景观资源，整体规划均围绕"江景资源"展开。

本地块住宅小区内部由10栋高层住宅、1栋多层住宅以及部分沿街商业组成。其中地块北面沿龙瑞路有一栋6层住宅，其首层为底商；东面沿江一侧为5栋一梯两户18层点式高层住宅，其建筑高度不超过60 m。

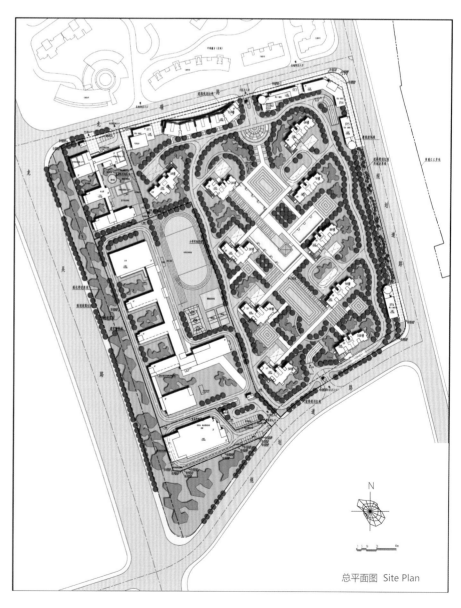

总平面图 Site Plan

MASTER AND MASTERPIECE | 名家名盘

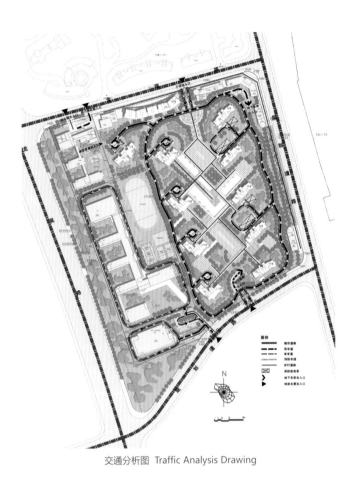

交通分析图 Traffic Analysis Drawing

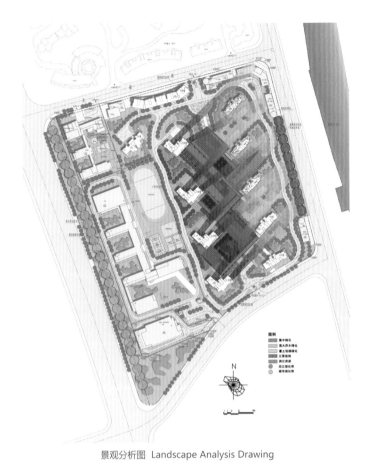

景观分析图 Landscape Analysis Drawing

建筑设计

项目立面采用新古典主义风格，以艺术装饰风格为基调，加以提炼、创新，在强调体积感挺拔沉着的基础上，强调时代感和创新性。整个建筑群体高低错落，天际线变化有致，立面上凹凸变化及窗的不同比例，表现了建筑外观变化和丰富的一面。

建筑上部以面砖饰面，下部基座饰以石材，色彩以暖调的浅褐色为基调，突出端庄、高雅的风范。入夜，顶部的天际线以泛光照明，形成优美的沿江夜景效果，从而塑造出一个非同凡响的现代化高级城市公寓的形象。

景观设计

城市外部绿化的借景：基地东西方向20 m，南北方向约400 m，有面积约8 000多平方米的绿化带，成为阻挡龙吴路噪声的天然屏障。黄浦江位于基地东侧，方案通过建筑的高度递减及角度借让，使80%以上的住户均能从不同的角度享受到优美的江景资源。

社区内部绿化体系的构建：按照江景资源的可视角度，基地内同样采用从东向西逐渐增高的景观层次，以减少高大植物对观景的影响。低矮的地库覆土绿化位于基地内东侧，而内部及西侧则以密植的绿化树木为主导，保证观景的同时，与适当的空间节点结合，构建起全区有疏有放、方便通达的景观绿化体系。

MASTER AND MASTERPIECE | 名家名盘

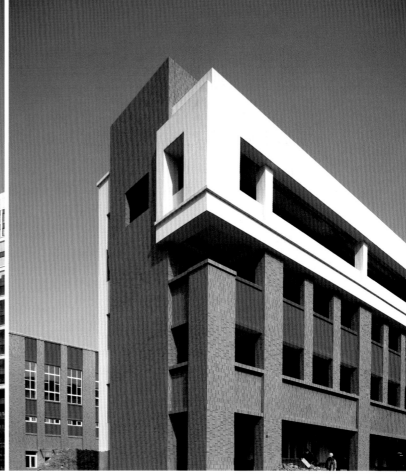

Profile

Located at Xuhui, Shanghai, the east side of the project is right adjacent to Huangpu River, the west is next to Long Wu Road and the north is opposite to Phase I with Long Rui Road lies between. Head to south about 1 km along Longwu Road from the project site, A20 Highway and Xupu Bridge will come into sight. The project serves as a residential community, and supporting public facilities, such as a 10-class kindergarten, a 30-class primary school, food market and shops, are located at the west of the site.

Planning and Layout

Under the principle of "efficient, rational, innovative", the project continues the planning context of the last phase, focuses on mining and developing environmental characteristics, forming a prominent feature. The Huangpu River is the most favorable landscape resources of the project, so the overall planning is all about "waterfront landscape resources".

It is a residential community that comprised of five high-rise buildings, one multi-storey building and some shopfronts along the street. On the north side along Long Rui Road, there is a six-storey residence with shopfronts on the ground floor. On the east side along the River, there are five point mode high-rise buildings no more than 60m, in which two occupants share one elevator in the same floor.

Architectural Design

Designers adopted neo-classic style in the facade and took Art Deco as the keynote, highlighting contemporary and innovation on the basis of emphasizing heaviness and volume. Entire building groups scatter high and low and the skyline changes regularly, different proportions of the windows and bumps on the facade express a rich and varied architectural appearance. Face bricks in the upper part of building, stone materials in the base and the sandy beige in warm tone highlight an elegant and exquisite style. When night falls, floodlighting in the skyline illuminates the waterfront, shaping an unusual image of modern high-class urban residence.

Landscape Design

View borrowing from the exterior greening: An 8,000m^2 greenbelt, 20 meters east-west and 400m north-south, is a natural barrier that keeps out the noise from Long Wu Road. As the east side is adjacent to Huangpu River, the height of the building is decreasing gradually and the view angles are adjusted properly to enable more than 80% of the occupants to enjoy the beautiful views from different angles.

Construction of greening system within the community: the landscape level increase gradually from east to west in accordance with the viewing angle. Low earthing greening is on the east side, while the close planting trees are on the west, which creates a convenient and accessible landscape system.

INTERVIEW | 专访

商业地产要整合好策划、营销与设计之间的关系
——访深圳市佰邦建筑设计顾问有限公司董事、总建筑师 王志强

■ 人物简介

- 王志强
  国家一级注册建筑师
  佰邦建筑设计公司（香港）创始合伙人
  深圳市佰邦建筑设计股份有限公司设计董事、总建筑师
  王志强先生专业设计大型公建、商业及城市综合体项目，对于交通类建筑项目设计工作颇有心得，同时对大型产业园区、开发区等城市新区建设及规划多有涉猎，完成过多个相关的优秀项目。

Profile:
Mr. Wang Zhiqiang
National first-class registered architect
Founding partner of P.B.A Architecture (Hong Kong)
Director & Chief Architect of P.B.A Architecture (Shenzhen)
Specializing in the design of large public buildings, commercial and urban complex projects and with rich experience in designing traffic buildings, large industrial parks and development zones, Mr. Wang Zhiqiang has created a number of related outstanding projects.

《新楼盘》：佰邦主张在满足项目功能和审美需求的同时突出其独创性，这种设计理念包含哪些方面的内容？在具体实施时要考虑哪些因素？

王志强：满足项目功能是职业素养，考虑审美需求是品味标准，而设计的独创性是信仰需求。不具备职业素养就不能在建筑设计界生存，品味标准的不同是服务不同市场人群的需要，而信仰需求，简单的说来，就是一种尊重，对土地的尊重，对环境的尊重，对人文的尊重。没有这种尊重，就会有功利性的设计，就会有破坏性的设计，就会留下许多将会被后人诟病的遗憾。

在面对每一个项目、每一个客户、每一块用地的时候，是否真正通过沟通、通过实地的走访、通过在用地上静静的思考，深刻的解读用地的性格，客户的需求，以及将来终端使用者的感受预期，才能将我们所主张的设计理念贯彻下去。这里，首先我们必须强调的是，充分的沟通和实实在在的换位思考是这一切的前提。因此，我们整合了综合服务平台，为的就是能够对每一个项目进行充分的沟通和服务。

《新楼盘》：将各类地产的策划、设计以及营销进行整合是你们的一大特色，它们各自在这个综合平台上扮演着什么样的角色？

王志强：必须要强调的，地产设计是一个复杂的设计过程，建筑设计在其中只是一个部分。要在地产设计中能够有效的沟通，精准的设计，必须要有各个部分的无间协同、通力合作，早在佰邦创立之初，公司就有着建立协同设计平台的思路，这个协同设计平台简单的来说，就是一个"特种部队"，成员单兵能力强，在各个领域能独当一面，整合起来又能够发挥整体实力。

这些年来，我们与多个地产服务公司进行协商合作，在项目配合的过程中判断双方的理念和配合度，找寻最佳的合作伙伴及合作关系。目前，我们已经形成了以地产策划、建筑设计、景观设计、营销推广为一体的地产服务平台，针对不同类型的项目，我们能够调配最优的服务资源，进行快速有效的沟通合作，在短时间内，提供客户多思路多方向的比较方案，并能够在熟知客户需求的前提下，协助客户进行关键阶段的策略判断。详细的来说，前期策划定位阶段，地产策划通过对市场的先期调研，通过科学的经济测算，提出合理的定位，有效的控制项目风险，这个阶段以策划为主，提思路提概念，设计为辅，推数据推沙盘，相辅相成，给到客户直观明确的项目模型进行判断。实际设计阶段，充分理解策划内容，运用专业设计手段，产生具体化合理性的方案设计，建筑设计、景观设计乃至室内设计的充分配合，减少施工过程中的错误率。营销推广阶段，全面熟知产品优势，产品亮点，能够精准营销、快速推广，实现最快的资金回报。

总的来说，整合后的地产服务平台，目的就是能够将地产服务从开始到完成都做到思想统一，无缝衔接，高度的完成度，精准的产品定位和投放。

《新楼盘》：就商业综合体而言，其前期的策划对于项目后期的建筑设计以及建成后的运营有何重要的意义？

王志强：商业综合体、办公综合体以及城市综合体，是目前我司操作较多的地产项目类型，这种类型的项目，特点在于综合，成败在于特定的产品，如商业、办公。因

此，正确的商业及办公产品的定位，关系到整个项目能否成立，这是项目首要解决的问题；其次，合理的产品配置，客观的产品分析，项目资金流的健康运转，从而降低项目运营风险；同时，合乎城市等级、地理位置和地域人文的产品形象，关系到项目对于城市的社会贡献。在将来的运营过程，则是在逐步的实现前期策划的意图的过程，使项目成为社会中有造血功能，良性循环的精神生活载体。

《新楼盘》：除了住宅和酒店建筑外，佰邦在商业综合体（商业建筑）等方面有很多非常优秀的项目，能不能挑选一个近期案例与大家分享一下？（可从策划、设计以及建筑特点等方面来介绍。）

王志强：商业建筑的成败要靠实际运营的成败来判断，因此没有个几年的跟踪回访，不能说就是成功的设计。我简单介绍一个"难点"项目的设计策略，在内地省份一个区级城市，有一个40多万平方米的商业综合体项目，商业面积达到12万平方米，办公面积14万平方米，余下为五星级酒店，产权式酒店，酒店式公寓等产品。项目难点为：内地城市，新开发地区，居住容量尚未形成，周边类似商业楼盘还有两个。设计难点为：用地面积9万平米，建筑覆盖率30%，商业面积必须做到12万平方米。这里我们简单的计算一下就可以得出，如果满铺覆盖率，12万平方米的商业要做到6层，在内地新区，三层以上的商业价值就很难保证，如何将商业业态处理好，商业做活，是项目的成败的关键。因此，我们协同策划公司、商业代理公司针对本项目进行了反复的推演，前后建立了十多个数据模型，最终，将本是缺点难点的覆盖率发展成为本项目的特点亮点，引入公园化情景街区的设计手法，保证在一期运营中打造与众不同的商业特征，强势的将本地块商业特点植入，配套以快打快销的产品策略，实现资金投入的快速回现，目前本项目正在实施过程中。

《新楼盘》：城市综合体与一般的公共建筑相比，在设计内容和形式上有哪些不同？

王志强：城市综合体在社会发展、城市建设、生活改善的共同作用下，已经成为我国城市建设和人民生活改变的重要组成部分，这点在高密度人居环境的城市尤为明显。城市综合体是建筑综合体的概念，主要特征是：1、内部包含了城市的公共空间；2、内部各部分的功能之间，有类似城市各功能之间的互补、共生关系；3、具有三种或者三种以上的能够产生效益的功能；4、项目构成中功能和形式高效的统一；5、强调城市、建筑、市政设施的综合发展。

城市综合体有助于城市化的发展水平以及城市经济和城市空间结构的调整。其协同效应可以节约城市用地，提高办事效率，创造低碳城市环境，改善城市生活空间，提高城市的生活质量，这些方面非一般性的公共建筑能够实现。

《新楼盘》：近年来，以商业综合体为代表的商业建筑开发非常火爆，你们是怎么样看待这种现象的？未来商业建筑的前景又是怎样的？

王志强：由于城市化发展的需要，以及国家对于地产项目宏观调控的决策下，目前以商业综合体为代表的商业项目数量较过去几年增加了许多，也证明这是一个发展的趋势使然，但是，缺乏统筹判定的商业模式导致重复性的商业项目的出现，为将来商业竞争带来了诸多隐患。目前匆匆上马的许多所谓的商业综合体并没有把商业作为主要成败的决定点来做研究分析，将盲目自信、缺乏市场判断的决策作为独有的商业运营思路强推，将会带来未来商业市场的恶性竞争，催生更多的社会矛盾，在这样的现实下，商业地产设计群体的整体成熟显得尤为迫切。若干年后，将出现一类能够精准判断、准确定位的商业地产设计机构，为城市更新、城市发展所需求的高标准商业项目提供服务。同时也将因目前的情况催生出一类为解决盲目上马而带来的垃圾商业项目困局的再生商业设计机构。届时，这两类机构都将更加广泛的进行商业地产设计平台的搭建，除去整合目前我们已经整合的相关设计服务机构外，还更广泛的整合商业资源、商家资源，更有效的为不同的商业项目类型服务。

Commercial Property Should Integrate the Relationship among Scheme, Marketing and Design Well

—— Wang Zhiqiang, Director & Chief Architect of P.B.A Architecture (Shenzhen)

New House: P.B.A Architecture proposed to highlight the originality when meeting the functional and aesthetic needs, what does it contain in this design concept? and what are the factors to be considered in the specific implementation?

Wang: Meeting project's function is professionalism and considering the aesthetic needs is the standard of taste, while pursuing design originality is a matter of demand in belief. No Professionalism no survival in the field of architectural design, different standards of taste is to service the needs of the different customers, and demand in belief, simply said, is a kind of respect, respect for the land, respect for the environment and the humanities. Without this respect, there will be utilitarian design, destructive design and even regrets that would be blamed by later generations.

In the face of every project, every client, every piece of land, only by real communication, field visits and deeply interpreting the character of land, the needs of customers and expected feeling of future end users, can we implement the design philosophy we proposed. What we should get to know in the first stage is that adequate communication and putting yourself into others' shoes are the premises. Therefore, we integrate a comprehensive service platform, just to get the sufficient information of each project and provide the best sincere service.

New House: It is a great characteristic of you to integrate scheme, marketing and design in all kinds of real estate, what roles did the three play in this comprehensive platform?

Wang: I have to emphasize that real estate design is a complex process, and architectural design is only a part of it. To achieve effective communication and accurate design, whole heartedly cooperation among every parts is needed. As early as the beginning of the creation of P.B.A Architecture, the company had the idea of establishing a collaborative design platform, simply speaking , it was a "special forces" with strong ability, not be able to work independently in various fields, but act as whole strength when integrate together.

In these years, we consulted and cooperated with a number of real estate services to find the best partners and cooperative relations through judging the degree of adaptability in cooperation. Currently, we have formed estate planning, architectural design, landscape design, marketing as one of the real estate services platform, for different types of projects, we are able to deploy the optimal service resources, rapid and effective communication and co-operation in a short period of time, to provide customers more alternative schemes, and to help customers determine in the critical phase under the premise of knowing them quite well. Specifically, in the planning and positioning stage, reasonable positioning has been put forward after scientific econometric measurement, which controls the project risks effectively. In design stage, scheme is fully understood, professional design techniques is used to generate specific reasonable design, architecture design , landscape design and interior design cooperate fully with each other to reduce the error rate in the construction process. In marketing stage, knowing the product advantages comprehensively keys to precise marketing and rapid promotion, which helps to achieve the fastest return on capital.

In general, the purpose of integrated real estate service platform is to be capable of being united from the beginning to the end.

New House: Just as in case of commercial complex, does the earlier planning have any significance for the later architectural design and the operation after completion?

Wang: Commercial complexes, office complexes and urban complexes are the major works of our practice that characterized by comprehensiveness, and the success or failure of these projects lies in the specific products, such as commercial, office. Therefore, the correct positioning of the commercial and office products, related to the found of the entire project, which is the most pressing issue of the project; Second, reasonable product configuration and objective analysis of the product concern to the healthy running of the project funds and can be able to reduce the operation risk of the project; in addition, product image in line with the city level, geographic location and geographical humanities relates to the project's social contribution for the city. Operation is a process that realizes the earlier planning, which makes the project a carrier with hematopoietic function and virtuous circle of spiritual life in the society.

New House: In addition to residential and hotel buildings, P.B.A Architecture has a number of good works on commercial building, can you share some recent cases with us in the aspects of planning, design and features?

Wang: The success of commercial building depends on the actual operation. You can't say it is a successful design without years of tracking return visit. I will briefly introduce the design strategy of a "difficult" project. There was a commercial complex project over 400,000 m² in an inland district-level city, of which the commercial area is more than 120,000 m², the office area is about 140,000 m², and the rest are for five-star hotel, property hotel, service hotel and so on. A newly developed zone in an inland city where the living capacity is far from being formed and there are two similar commercial properties around is what the difficult about the project itself, while creating 90,000 m² floor area, 30% building coverage and 120,000 m² commercial area is where the difficult lies in the design. Therefore, we did the research again and again under the cooperation with scheming company and commercial agency, after several models had been discussed, we turned the shortcomings of coverage rate into characteristics of the project, which will soon make the money back.

New House: What are the differences between urban complex and public building in terms of design content and design form?

Wang: Under the interaction of social development, urban construction and improved living standard, urban complex has become an important part of China's urban construction and changes in people's lives, and this is particularly evident in the high-density living environment. Urban complex is the concept of the building complex, the main features are: 1, contained the city's public spaces; 2, there is certain complementary relationship between the internal functions of each part just like that among the city's various functions; 3, have three or more functions that can produce benefit; 4, efficient integration of functions and forms; 5, highlight on the comprehensive development of the city, construction and municipal facilities.

Urban complexes contribute to the development level of urbanization and the urban economy and the adjustment of urban spatial structure. Their synergistic effect can save urban land use, improve efficiency, and create a low-carbon urban environment, improve the urban living space and the living quality. That's what couldn't be achieved by common public buildings.

New House: In recent years, commercial building represented by urban complex are very popular, what do you think of this phenomenon? And what's the future prospect for commercial buildings?

Wang: Due to the needs of the development of urbanization and decisions of national macro-control regarding to real estate projects, the number of commercial projects, represented by commercial complexes, has increased a lot over the past few years, which proves that it is a development trend. However, business model lack of overall planning and co-ordination leads to the emergence of the repeatable commercial projects, bringing a lot of hidden troubles for the future commercial competition. At the moment, some of the hastily implemented so-called commercial complexes do not take commerce as the major point to do the research, but tend to decisions lack of market judgments, which will brings future commercial vicious competition and give birth to more social conflicts. Under this circumstance, the overall advance of the commercial real estate design is particularly urgent. A few years later, commercial real estate design agencies with accurate judgment and accurate positioning will meet the needs of the high-standard commercial projects for urban renewal, urban development. In addition, some renewable commercial design institutions will come into being to cope with the problems that caused by the so-called commercial projects. By then, both of them will create commercial real estate design platforms more extensively, beside integrate the related design agencies we have integrated, they will integrate business resources even further, providing services for different commercial projects more effectively.

INTERVIEW | 专访

"我们憧憬的未来"——诗意栖居

——访广州山水比德景观设计有限公司执行董事、总设计师 孙虎

> "人应该诗意地活在这片土地上。这是人类的一种追求一种理想。"
> ——[法·数学家、思想家]布莱茨·帕斯卡尔
>
> "人类社会面临资源枯竭、环境恶化等重大挑战。今天的世界，已经没有新的大陆和绿洲可被发现，保护资源环境、实现永续发展是我们唯一的选择。"在今年6月20日于巴西里约热内卢召开的联合国可持续发展大会上，温家宝总理大声呼吁。
>
> 迫于显见的危机，与会各国达成了一致的行动框架，也即"我们憧憬的未来"。
>
> 全球在行动！
>
> 作为景观设计行业的一家新锐创意机构，广州山水比德景观设计有限公司在此一方面，可谓不遗余力。从其企业使命——"让人类诗意栖居"，或可管窥。
>
> "诗意栖居，是人类的共同居住理想，是大家所共同的憧憬。"山水比德执行董事兼总设计师孙虎指出，诗意无国界，以艺术取自人与自然、社会间的平衡共生。

■ 人物简介

孙虎先生

山水比德景观规划创始人兼首席设计师。首届羊城青年设计师大赛金奖得主，先后荣获过"中国国际园林花卉博览会综合大奖及设计金奖"、"青工科技成果奖"、"香港皇家园林博览会金奖"、"广州园林博览会特别大奖"以及"2007年度中国100位杰出建筑规划（景观园林）设计师"等荣誉。曾先后主持和参与过近百个国际国内重要景观规划和景观设计项目，多次获得重大国际和国内奖项，设计项目包括市政景观规划设计和地产类景观规划等多种类型，在景观的创意性设计方面有着独特的见解。

Profile:

Sun Hu: the founder and chief designer of SPI Landscape Design Ltd, gold medal winner in the First Guangzhou Young Designer Match, goal medal winner of comprehensiveness and design match in China International Garden Expo, one of the top 100 architectural designers (landscape and gardens) in China 2007 etc. He has successively in charge of and taken part in the planning and designing of early one hundred important landscape projects at home and abroad, obtained great international and domestic rewards. His projects had varied from municipal landscape design to real estate landscape design etc, and he shows his unique insight in the creative design of landscape.

设计，为生态系统提供条件

《新楼盘》：资源枯竭、环境恶化，以及城市化的过快发展，带来一系列的城市病。您认为，面对种种困境，景观设计师应当如何作为？

孙虎：工业革命给人类带来前所未有成就的同时，却也使得人类的自信心受到前所未有的膨胀，征服自然、改造自然，最终却也狠狠地将自然破坏，将人、社会与自然间的平衡、共生关系打碎。

"亡羊补牢，犹未晚也"，重拾山水，恢复人、社会与自然间的平衡共生关系，成了当前的重头戏。基于上述的时代语境，"可持续发展"也便成了当前语境下的时代主题。

此一点，景观设计师，亦不例外——即以"可持续发展"为主题，探索当前语境下的景观创意之路，并以之解决当前困境。

《新楼盘》：作为景观设计行业的资深设计师，您认为景观设计在环境问题中所能起到的作用是怎样的？

孙虎：一直以来，我们很多景观设计师都在玩造型、玩花样，为打造项目的噱头而设计。然而，前不久的一次法国景观考察，却对我触动颇深：景观是为自然提供繁衍生息的可能，设计是为生态系统提供条件。

而景观设计师，则应通过对于自然的接触以及对场地的感知去做设计，用情感创造诗意空间。设计师只有深度了解场地及当地文脉，才能确切传达出场地精神以及培养出设计师对场地、空间的情感，诗意才能流露，生态与人、社会的平衡也才能实现。

玩造型，做噱头，浮躁的设计对于环境问题而言，于事无补；只有真正沉下心来做设计，理解场地，熟悉文脉，培养情感，诗意有成，生态的平衡也水到渠成。

这时，景观设计也才真正发挥其本质的作用。当然，国外是一步步发展到这个阶段，内地的景观设计也不例外，需要时间。

栖居，景观赋存诗意

《新楼盘》：据了解，山水比德一直以"让人类诗意栖居"作为自己的设计使命。那么，这是否出于上面所述的原因，也即以景观设计来探索可持续性？

孙虎：刚说过，"可持续发展"的时代主题其实源于当下的语境，也即在人、社会与自然间的平衡、共生关系遭到破坏的情境下提出的。

然而景观设计虽然可以归结为工业革命以及城

市化等现代运动的产物，但是追根溯源时，我们却可以发现，早在工业革命之前的园林艺术便已然有了景观设计的雏形。由此一点，可以看出，景观设计的需求由来已久，并不止于当前语境。

那么，我们就反思：景观设计的本来价值在哪里？或许，抛开当下语境，我们便可深层次了解。无论是景观也罢，还是园林，无疑都是人类活动的载体、栖居的环境。但人类却一直没停止过对此的研究、设计以期不断提升，那便是因为这个环境并未达到人类栖居的梦想。

因此，我们进一步反思：人类所憧憬的栖居环境是怎样的？

最终，在中西方的文化与生活比较中，我们提炼出一个共通的词汇——"诗意"。抛开语境，往往才能发现事物的本真。所以，我们所提出的"让人类诗意的栖居"，并非是源于当下的语境，而是源于人类对于栖居环境的最高要求与探索之上。

《新楼盘》：那么，"诗意栖居"的景观主张对于解决当下环境困境，又有着怎样的助益？

孙虎：经过我们多年对于国内外文化与生态的研究、探讨，加之山水比德在景观设计领域的沉淀与发展，我们对"诗意栖居"有了进一步的认知与主张。

从对象来说，诗意无国界，是人类共同的栖居梦想与追求；从内容来说，诗意是共生，是人、自然与艺术的平衡性共生；从性质来说，诗意，有个性，陶渊明是"悠然见南山"、海子是"面朝大海，春暖花开"……各有不同，我们便是定制属于客户需要的诗意；从本质来说，诗意是生活的态度、方式，我们设计的是景观，更是引领健康、舒适的生活时尚；从路径来说，诗意是交互，是艺术审美、文化沉淀以及科学理性等交互路径下的景观新界面……

可见，我们的"诗意栖居"主张不仅仅将追求平衡、共生的"可持续发展"作为题中之义，更是有着很大的发展与升华。

诗意，现代的随意

《新楼盘》：请您结合山水比德以往的案例谈一下什么样的景观才具诗意？

孙虎：我们可以从大一山庄项目的景观设计来进行解读。

大一山庄北区项目完成之后，我们对此进行了回访与调研，其中就发现，许多业主都会有一个共同的误解：项目场地原本便是绿意盎然的林海，景观设计不过是就势借势，最终形成了"房流于林影"的效果。而事实是，场地原本一片荒芜，仅几颗树木寥落。

何以有这样的反差？关键就在于设计方案中对于"现代的随意性"的营造。

在设计方案中，"水、月、层、云、林、艺、奇、然"的景观意向并非天马行空，而是我们根据场地精神所确立的"因地制宜"；植物搭配上的选择，以及石块运用，亦是考究于当地资源，而做出的平衡性方案……凡此种种，都不见设计上的人工痕迹，而是觉得"本来天成"，俨然是自然的随意而生。

为什么我们要追求这样一种现代的随意性景观？因为诗意本源自然。无论中外的文学中，还是艺术再现中，我们发现那些人工痕迹明显的景色，如修剪整齐的人工草坪、花卉、光滑的河岸是没有诗情画意的，是难以入画入诗的；而唯有那些似"粗糙"却丰富，似随意却有法则意义的自然形态的植被、水域、石垒等方有入画的期许，也才能为人类提供富有诗意的感知与体验。

在此一点上，大一山庄分外明显，也因此让许多身临其间的业主、游客，流连忘返，诗意本天成！

《新楼盘》：对于"诗意栖居"，这些年，山水比德在项目上有着怎样的具体追求？

孙虎：我们认为，诗意三层：画意其外，触景生情；含蓄其中，逗人赋诗；沉耽其中，忘乎所以。在设计工作的实践层面，我们要求比德出品必有诗意流露，犹如画意其中；然后努力赋景观以诗意引力，人置于其中，必有作诗冲动，我们认为每个人心中都有一位诗人，关键在于外界引力是否到临界点；最后，我们所要达到的是，人过境之时，忘乎所以，以至于流连忘返，诗情也罢，作诗冲动也罢，都已抛之脑后，神随景动，一切恍惚。

可以说，从事务所起步时的诗意浅尝，到现在的高层次升华，我们在项目上诗意追求正在不断地提升与加强。可以说，大一山庄充分体现了我们在诗意上的追求与实现。

广州大一山庄

广州大一山庄

INTERVIEW | 专访

The Future We Look Forward to — The Poetic Inhabitation
—— An Interview with Sun Hu, Managing Director and Principal Designer of Guangzhou SPI Landscape Design Ltd

"Human deserves poetic inhabitation on the land, which is our pursueit and wish."
—by Blaise Pascal, French mathematician, ideologist

"Human beings are facing severe changelings like the exhaustion of resources, deterioration of environment. Given no discovery of any more new landmass or oasis, the protection of resources and environment for sustainable development comes to our only choice.", appealed by Premier Wen at the UN Conference on Sustainable Development in June 20th this year in Rio de Janeiro, Brazil.

Threatened by the obvious crisis, the attendee countries has reached action framework in unanimity — "the future we look forward to". Actions around the world!

As a new, vigorous and creative company majoring in landscape design, Guangzhou SPI Landscape Design Ltd has taken every effort to work for that, which could be perceived from the enterprise mission— building the poetic inhabitation.

"Poetic inhabitation is the people's shared inhabitation wish, and is what all of us look forward to." Sun Hu has indicated that it is boundless, obtaining the co-existence in balance between human, nature and society through art.

Design: creating and offering conditions for ecology

New House: A series of city diseases come out of the exhaustion of resources, deterioration of environment and the speeding urbanization. In your opinion, what should a landscape designer do in face of various difficulties?

Sun Hu: For one thing, the industrialization brings us unprecedented achievement; for another, it inflates people's confidence unprecedentedly. People are eager to conquer and transform the nature, which will end up in great damage to the nature and the balance between human, society and the nature.

However, it will never be too late to make it up. It will be a key part to restore the coexistence and balance between human, society and nature. Given for the mentioned conditions above, "sustainable development" consequently becomes the theme of the time in the current context. From this point, landscape designers are no exception---centralized with "sustainable development", exploring the creative landscape design and the solution to the predicament.

New House: As an experienced landscape designer, what is your opinion about the landscape design's role playing in the environment problem?

Sun Hu: Many of the landscape design have been no more than showoff and stunt for a long time. However, a France landscape tour not long ago stirred up my feelings deep inside. Landscape is supposed to guarantee the nature propagation and the design is serving for the ecological system.

For the landscape designers, they are supposed to work with their contact with the nature and their perception about the sites, to build a poetic space with feelings. Only after a further study of the site and native context, could a designer present the site spirit, develop a feeling towards the site and space, which leads to the poetic revelation and the realization of the balance between ecology, human and society. Showiness and stunt has nothing helpful to solve the environment problem, only the real design after studying the site and context, develop the poetic feelings could lead to the ecology balance.

That is when the landscape design plays its real role. The development abroad is moved step by step, and so no exception at home, which requires time.

Inhabitation, poetry in the landscape

New House: As we learn that SPI has kept holding "building the poetic inhabitation" as your design mission. Is your company exploring the sustainability through landscape design out of the reason mentioned above?

Sun Hu: Just as I have said, the theme of the time "sustainable development" results from the current context, and it is brought up in a condition that the balance and coexistence between human, society and nature is suffering from damage.

Though landscape design could be ascribed to the production of modern movements such as industrial revolution and urbanization etc, we could trace back and find that it has been in embryo in garden arts long before the industrial revolution. From that point we could affirm that the demand for landscape design has been a long history and more than the current context.

It thus leads us into contemplation: what's the original value of landscape design? We may have a further understanding when standing far away from the current context. Both landscape and gardens are the supporters for human activities and inhabitation environment, and people keep studying and designing with the hope that it will improve continuously, since the environment has far beyond what we dream for inhabitation.

Then we will take a further thinking: what kind of inhabitation environment do people look forward to?

Finally we abstract a common word out of the comparison between the culture and living in east and west--- "poetic". An essence could be discovered when the context is set aside. So the idea of "building the poetic inhabitation" did not come out of the current context but of the highest demand and exploration for human inhabitation environment.

New House: What positive role will the landscape idea "building the poetic inhabitation" play in solving the current environmental predicament?

Sun Hu: After years of studying and exploration on culture and ecology at home and abroad, in addition to SPI's experience and development on landscape design, we have had a further realization and perception about "poetic inhabitation".

In view of the target, poetic inhabitation is a common pursuit of human, without boundary; in view of the content, it is a coexistence and balance between human, nature and art; in view of the character, it could be of individuality, Tao Yuanming likes to "glance at Nanshan leisurely and carefree", Hai Zi prefers to "facing the sea with spring blossoms" … it has various types and what we do is to present to our customers their exclusive poetic way. In view of the essence, it refers to lifestyles and we design the landscape with intention to take a lead in a healthy and ease lifestyle fashion. In view of routes, it is an interaction, a new landscape interface under the art esthetic, culture precipitation and scientific reason etc.

Our idea of "poetic inhabitation" is in pursuit of not only the "sustainable development" as a means of maintaining the balance and coexistence, but also more development and improvement.

Poetic with modern random

New House: Could you share with us your opinion on poetic landscape in reference to the cases by SPI?

Sun Hu: We could take the landscape design of Dayi Mountain Villa as an example. After the completion of the north zone, we paid a return visit and research, and then found a common misunderstanding most proprietors hold: the site was an immense forest before the project started, ad the landscape design was supposed to match up the surrounding environments and create the forest shadow around the architectures. However, the site becomes desolate with few trees.

What leads to such a contrast? The key lies on the building of "modern random" in the project planning.

In the planning, landscape elements of "water, moon, layer, cloud, forest, art, unique and nature" are not built without base, they are designed according to the site condition; the selection of plants, and the application of stone and rocks are considered to maintain the balance of the local resources. All of these are presenting a sense of random and nature and minimum trace of artificial design.

Poetry comes from nature; this is why we require for landscape of modern random. Landscape of artificial trace could be found in either literatures at home and abroad, or art recurring, such as the manicured lawns, flowers and glossy river banks; they are not poetic. Only the plants, waters and rocks with rough but rich nature forms, random but of regulation significance could be placed in the "picture", could provide to human poetic sense and experience

Dayi Mountain Villa obviously gains a great achievement from this point, which attracts the proprietors and visitors with its natural poetry.

New House: What's the exact requirement for poetic inhabitation in SPI's project for years?

Sun Hu: We consider the poetry as 3 levels: the sceneries will excite the emotions, the implication will arouse the poetic feelings and the further reading will lead to a fascination. In the practical level of the design, the landscapes designed by SPI are required to be poetic like picturesque sceneries and will arouse people's interest of poems; we do believe that every man is a poet and the attractions outside matters when it comes to a key point. Finally what we require is people's fascination of the landscape which makes them forget themselves.

From a poetic try in the first early time, to the sublimation in higher level, we have been improving and developing our pursuit of poetry. Dayi Mountain Villa is such a project definitely demonstrating our pursuit and achievement.

NEW LANDSCAPE | 新景观

项目地点：美国旧金山
景观设计：Andrea Cochran 景观公司

ECO RESIDENTIAL LANDSCAPE

Drs Julian and Raye Richardson Apartments

生态节能、绿色高效的住区景观——旧金山理查森公寓

项目地点：美国旧金山
景观设计：Andrea Cochran 景观公司

Location: San Francisco, America
Landscape Design: Andrea Cochran Landscape Architecture; San Francisco

理查森公寓是旧金山市低收入居民保障房，共有120个设施完备的房间和配套服务，包括：咨询、医疗、就业培训和提供就业机会。公寓景观项目包括有：一条绿化通道、一个中央庭院及公寓房顶。项目建设就地取材，项目设计也使景观植物充分吸收来自渗水路面，楼内小园及楼顶的雨水。

设计整体策略是最大限度地利用雨水渗透到路面下的土壤，这对旧金山的下水道和雨水系统是非常重要的。设计中的街景，渗水路面砖和蓄水小园，不仅美化了公共空间，而且有效减少地面雨水流失，增加地下水补给。对街道的植物都大有好处。

楼内小园中等距安装有下水小装置，让雨水渗透到下面碎石蓄水地。溢出到地面的水通过小坡给园内棕榈树和蕨类植物吸收。室内的排水管系统也将楼内的水送到碎石蓄水地。屋顶的绿化植物，可观赏也可食用，以提高雨水吸取，减少城市大气含水量。

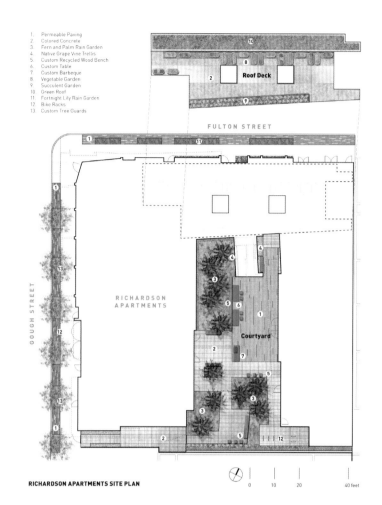

RICHARDSON APARTMENTS SITE PLAN

1. Permeable Paving
2. Colored Concrete
3. Fern and Palm Rain Garden
4. Native Grape Vine Trellis
5. Custom Recycled Wood Bench
6. Custom Table
7. Custom Barbeque
8. Vegetable Garden
9. Succulent Garden
10. Green Roof
11. Fortnight Lily Rain Garden
12. Bike Racks
13. Custom Tree Guards

NEW LANDSCAPE | 新景观

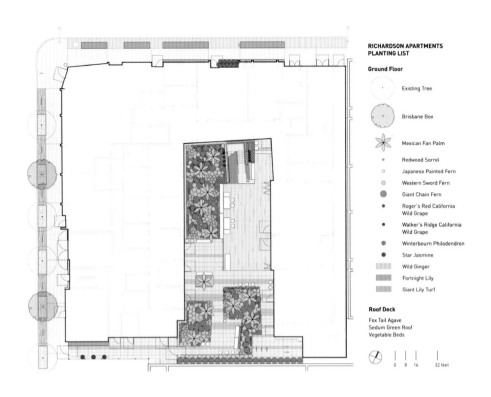

The Drs. Julian and Raye Richardson Apartments provide a dignified home for formerly homeless residents with 120 fully-equipped studios and supportive services, including counseling, medical care, job-training, and employment opportunities. The landscape design encompasses a streetscape, a central courtyard, and a roof deck—all fully-accessible with custom furnishings. The project uses local materials and offers a multi-faceted stormwater management with permeable paving over a gravel infiltration system, rain gardens, and a green roof.

An overall site strategy was developed to contain and maximize the infiltration of rainwater into the sandy site soils, which is critical for a city with a combined sewer and stormwater system. The innovative streetscape includes permeable pavers and rain gardens, which improve the aesthetics

of the public realm while reducing site runoff and allowing for groundwater recharge. Existing and newly planted trees along Gough Street benefit from additional water draining into the site soils.

Within the courtyard, spacers are installed between standard unit pavers to allow stormwater to permeate into the gravel retention basin below. Overflow and other paving areas are sloped into rain gardens, planted with palm trees and ferns. A perimeter drain also routes water away from the building and into the gravel retention system. On the roof, green roofs and planters, both ornamental and edible, also help to capture water that would otherwise enter the city storm system.

ECOLOGICAL RESIDENCE WITH HUMANISM AND HIGH QUALITY

Huidong International Garden

生态、人文气息的高品位心灵居所——广州汇东国际花园

项目地点：中国广东省广州市
开 发 商：广州市合汇房地产有限公司
景观设计：广州市柏澳景观设计有限公司
主笔设计师：徐农思
占地面积：32 500 m²
建筑面积：115 700 m²
容 积 率：3.00
绿 化 率：45.5%

Location: Guangzhou, Guangdong, China
Developer: Guangzhou He Hui Real Estate Co., Ltd.
Architectural Design: Guangzhou Bo'ao Landscape Design Co., Ltd.
Chief Designer: Xu Nongsi
Land Area: 32,500 m²
Floor Area: 115,700 m²
Plot Ratio: 3.00
Greening Ratio: 45.5%

　　汇东国际花园地块位于广州增城新塘镇。该项目与城市主干道、城市规划绿地相邻，位于广州市外环线——广深高速公路以东，广深高铁以南，相邻107国道。交通便利，自然环境优美。本社区规划用地面积32 500 m²，绿地总面积12 877 m²，绿地率高达45.5%，建筑为现代简约的中式风格。

　　汇东国际花园的景观设计，讲究整体风格为现代简约，低调中见奢华，丰富中见稳重，充分体现整个景观现代简约的中式风情，现代中式风格情调与建筑的宁静、轻巧神韵相互融合；传承发扬中国的人居智慧，创造精品高档生活风格，营造具有生态、人文气息的高品质心灵居所。

　　理念设计：新塘厚重的文化底蕴积淀着无数的沙和贝，新塘古时建筑多采用五间三进院落四合院式布局，穿斗式抬梁结构，屋顶为硬山顶。锅耳山墙，灰塑脊，炉灰筒瓦，属于较具特色的岭南建筑风格之一。现代中式风格，是植根于新塘岭南建筑文化之上，并巧妙结合东西方文化的一种新思路，设计师可以在保持岭南风格不同的形式展现。在设计中运用平面构成的原理和中式园林法则，以中式特有的有机图形构架为蓝本，通过简洁的直线、曲线和折线相结合，将整个居住区环境进行合理的分割，创造出更加明确、合理的功能空间划分。

　　销售中心位于该项目东北角，与107国道相连接，是该项目最主要的对外展示景点之一。销售中心前面富有浓厚中国风情的叠级水景，形成较好的标志性作用将人们的视线引至社区。传承中国传统的造林手法，通过采用借景、框景、遮景等造景手法，做到疏可走马密不透风，将整个社区划分为若干个景观空间的过渡，并通过线条的有机分割，形成一个完整的社区环境空间。

　　道路、树阵、花海等运用软景和硬景的完美结合，恰如其分的围合和分割了各个功能空间，使得体块既有分隔，又相互渗透。无论是古朴古香的入口大门，还是传统中式与现代相结合的艺术雕塑、景墙、蓝色的水景等等都散发出空间的强烈现代中式艺术气息。

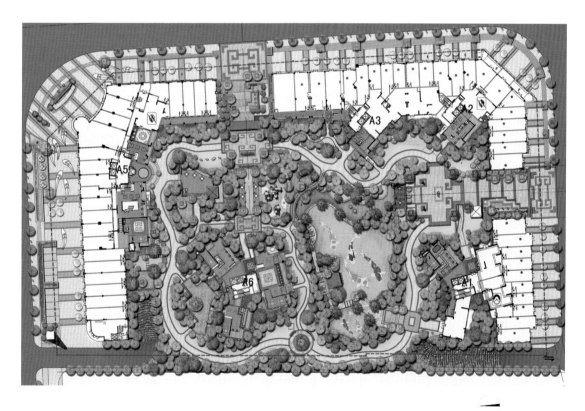

总平面图 Site Plan

NEW LANDSCAPE | 新景观

The project locates in Xintang Town, Zengcheng Guangzhou. It is adjacent to the city main stem and the city greenbelt, situating in the Guangzhou Outer Ring which is in the east of Guang-Shen expressway and the north of the Guang-Shen high-speed rail and next to the national road 107. Besides, it has convenient traffic and excellent physical environment. The project has a land area of 32,500 m^2 and a total greening area of 12,877 m^2 with a high greening ratio of 45.5%.

The landscape design of Huidong International Garden strives for modern, simple, modest and luxury to reflect the Chinese modern landscape design. It carries on Chinese people's wisdom and creates boutique high-end living style of ecological high-quality residence.

Design Concept: The rich culture heritage of Xintang leaves many sand and shells. In the ancient time, the buildings usually apply the layout of five three-courtyard quadrangles, bracket-crossing post-and-lintel structure

and hard roof. The gable wall, ash plastic ridge and imbrex are the vivid portrayal of the Lingnan architecture style. However, modern Chinese architecture style is rooted in the Lingnan architectural culture and skillfully combines the eastern and western architectural culture. For the design, the project uses the principles of plan formation and Chinese garden rule, using the neat straight line, curve and fold line together to make a reasonable segmentation on the whole living area thus creating more explicit and reasonable functional space.

The sales center lies in the northeast side, connecting the National Road 107, which is one of the main display spots. It has marked stacked waterscape to call people's attention in front. Basing on the Chinese traditional afforestation technique such as view borrowing, enframed scenery and view covering, the whole community can be divided into several sequential landscape spaces to form an unbroken environment space.

Under the perfect combination of soft set and hard set, the road, trees and flower sea appropriately combine and divide every functional space. Whether the primitive entrance door or the art sculpture, sign wall and blue waterscape shows out the strong modern Chinese art breath.

MOON WATER CLOUD, POETRY WORLD | Guangzhou Da Yi Mountain Villa

水月云天 诗意人间 —— 广州大一山庄

项目地点：	中国广东省广州市
开 发 商：	广东中力投资有限公司
景观设计：	广州山水比德景观设计有限公司
规划面积：	184 000 m²
景观面积：	118 000 m²

Location: Guangzhou, Guangdong, China
Developer: Guangdong Zhongli Investment Company
Landscape Design: Guangzhou Sansui Bidder Landscape Design Co., Ltd
Planning Area: 184, 000 m²
Landscape Area: 118, 000 m²

　　项目名曰"大一",意在追求天人合一,天道归一之境界——大之加一谓之"天",大之减一谓之"人"。因此,在项目的定位上,便有"此景只应天上有,落入人间为仙境——诗意人间"的立意。

　　结合项目定位、甲方需求以及场地资源等情况,设计方案也是呼之欲出。经过思想的碰撞,心灵的感悟,承袭项目之初"水、月、层、云"的理念和"天法道,道法自然"的原则,感悟"水、月、层、云、林、艺、奇、然"的景观意向,本次方案以《桃花源记》为引子,在意境、文化、精神上进行完美的升华,形成了本次项目规划的理念:水月云天•桃花源——天上人间。

　　在景观水体的总体布局中,设计师在竖向和平面上都进行了详尽的考究:结合项目基址的山形水势,在群山之间规划一条瀑布,形成银河落九天之势;瀑布之水流向整个社区,整个社区之水系即为玉带,用"玉带船说"的形式来加强社区内部的互动交流;组团之间的集中水面形成"七星拱月"之势!使得整个的水体系统形成一个有机的整体。

　　月,代表的是一种境界,一个情结!本案设计中将月的概念和意境引入到景观之中,就是要让家的概念深入人心,让"大一山庄"成为业主心里最盼望回归的家;云,层云叠影,形色万千、无穷想象。景观设计中通过雾泉的处理方式为社区的整个意境营造这种氛围;天,意即天人合一、天上人间。"大一山庄"蕴含着天道归一之境界,"大一"之中更是包容万象,暗喻天人合一之境界,是人间的仙境。

　　整个项目的设计原则,从理性与美学双重层面提出:前瞻性原则,基于现状、高起点、高标准,力求自然与艺术的统一、环境与人文的统一;生态可持续发展原则,应考虑协调当前与未来的平衡,正确处理自然资源保护与开发建设的关系,科学规划、合理布局;独特性以及人文性相结合的原则,以艺术化、文学化、心灵化、自然生态化、使园林渗透、充盈着诗意或者文心,从有限的建筑传达出抒情的无限想像;以人为本原则,意即强调不同层次、不同年龄阶段人群的参与性、与水的亲和性、休闲性与娱乐性。

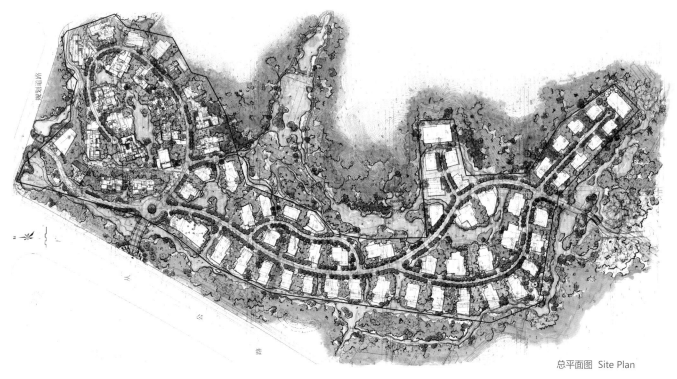

总平面图 Site Plan

NEW LANDSCAPE | 新景观

The Project named "Da Yi" means to pursue a state that harmony between man and nature, unification of natural law——Add Yi to Da means Tian, subtract Yi of Da means Ren. Hence, the positioning of this project is landscape should only be in heaven, fall into the world is fairyland—the concept of Poetry world.

The design proposal combined orientation of this project with client's demand and site resource. The design concept originated from Peach-Blossom Spring, sublimates artistic conception, culture and spirit, and formed the idea of this project planning which is Moon Water Cloud, the Peach Garden——Heaven on Earth.

The designers investigated the vertical and horizontal plane elaborately in the process of general layout of landscape: a waterfall was planned among mountains through uniting the river and mountain of the site, which formed the momentum that the Silver River fell down from azure sky. The waterfall flowed through the whole community and the drainage system in the community is jade belt, to enhance the interaction within community via "Jade belt legend". The central water in this housing cluster developed into a pattern of "Seven-stars with Moon", made the whole water system developed into an organic entity.

Moon stands for a kind of realm, or complex in China! The design of this project brought in the concept and complex of moon for the sake of interiorizing the concept of home, to make Da Yi Mountain Villa be a home that the owner long for most. The features of clouds are thousands of shape, boundless imagination. The designers built an atmosphere of cloud through dealing with fog spring in designing the landscape. Tian means harmony between man and nature, heaven on earth in this project. Da Yi Mountain Villa contained a realm that natural law united, the connotation of "Da Yi" is all-inclusive, with the metaphor of a state that harmony between man and nature is the heaven on earth.

The design principle of the project pointed out form double level of sense aesthetics: the perspective principle is based on current situation, high starting point and high standard, strived to unification of nature and art, integration of environment and humanity. Ecological sustainable development principle should consider the balance between present and future, manage the relationship between natural resource protection and development and construction, plan scientifically, distribute rationally; the principle of peculiarity and humanity united, to let this garden be filled with poetry by those mode of artistry, literacy, spiritualization and nature ecologicalization, transmit infinite lyric imagination from finite buildings; The principle of people-oriented emphasized on participation of people from different level, different age grades, the affinity, recreation and entertainment with water.

NEW LANDSCAPE | 新景观

TRANQUIL SIMPLE NOBLE ELEGANT

| Changsha Vanke Golden Mansion Phase One and Two

静谧 简约 高贵 典雅——长沙万科金域华府（一、二期）

项目地点：中国湖南省长沙市
开 发 商：万科长沙市领域房地产开发有限公司
景观设计：深圳市雅蓝图景观工程设计有限公司

Location: Changsha, Hunan, China
Developer: Vanke Changsha Lingyu Real Estate Development Co., Ltd.
Landscape Design: Shenzhen ALT Landscape Engineering and Design Co., Ltd.

长沙万科金域华府项目位于长沙市Ⅱ类区——长沙市雨花区，武广新城的核心区域，二环外，地块南侧为城市干道香樟东路，紧临雨花区政府和雨花市民广场，西侧为长沙市南北主干道的万家丽路，北靠湖南联通总部。

一期景观设计要点

小区主入口景观：小区"门户"。通过电梯进入到小区，此处设计了跌水景观和入口特色铺地，以及泳池外侧的水景幕墙，营造出一种与外部繁华商业空间截然相反的静谧舒适且豪华大气的居住空间。

宅间景观：户户临景，移步异景。既需要开放的绿地，也需要有属于个人的私密空间。室外运动场所以及架空层空间的合理利用，动静结合，为住户提供面积充足，设施齐备的硬质及软质场地，以吸引用户走出房屋，加入公共活动，以增进住户间交往。

中心花园景观：小区"客厅"。一期中心花园景观设计延续入口跌水景观水系，形成一个集中水域。动态跌水与平静水面的完美结合，喷水雕塑与景观雕塑的合理布局，增强对景的同时又富于变化，沿水区有亲水木平台，特色景观亭更为住户创造一个与水互动的空间。

二期景观设计要点

1、框架的形成：通过运用不同粗细的艺术线条，有规律的进行碰撞和交织。在线与线交和碰撞中形成的面与面之间通过植物、铺装、水体等不同的图底关系，丰富和充实着整个框架结构。

2、细节的考究：玻璃、金属、木材、石材等现代的装饰元素在细节上的运用是设计所要表达的另一种视觉和触觉的体验。通过这些材料的搭配和尺寸的考究，营造出简约、朴实又不失高贵典雅的景观气质。 3、不同层面的景观感受：以人为本的景观理念指导着我们始终将不同的人们对景观的感受放在设计的首位。通过不同材料的围合，生活小品的设计，来制造景观氛围。空间的思想，意在通过空间来启发不同年龄、不同背景、不同身份的人们对生活、对社会的深层次的思考和认知。

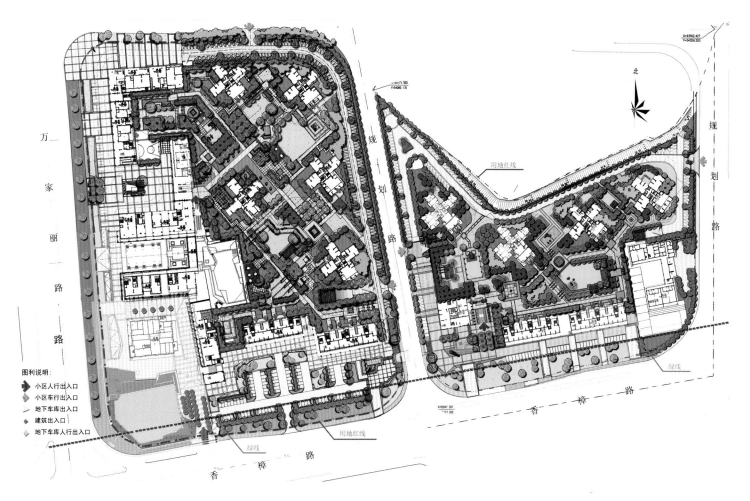

总平面图 Site Plan

NEW LANDSCAPE | 新景观

Changsha Vanke Golden Mansion is located in the Second Class area of Changsha, Yuhua District, central area of Wuguang New Town. It's sitting outside the second ring, with Xiangzhang East Road-main urban line on the south, Wanjiali Road-the main south-north line on the west and Hunan Unicom Headquarter on the north. The site is also closely next to District Government of Yuhua and Yuhua People's Square.

Key Point for Landscape Phase 1

Landscape at the main entrance: the gateway of the residence.

Here features water drop landscape, characteristic floor and the waterscape curtain wall surrounding the pool to create a tranquil, comfortable but luxury residential space opposite with the outside commercial space.

Between home landscape: Every family gets access to landscape; landscape changes with moving of steps. Residents in community needs not only open green land, but also private space which belongs only to oneself. The proper utilization of outdoor sports ground and overhead space, a combination of tranquil and dynamic, provides both hard and soft ground with full space and complete facilities to attract residents to walk out of rooms and join the public.

Central garden: the guest-receiving room of the community.

The central garden design of Phase One joins the water drop landscape

at the entrance to form a centralized system. The perfect combination of dynamic drop and tranquil water and the proper arrangement of spring sculpture and landscape sculpture strengthen comparisons and changes. Along the water sets close water wooden platform—specialty sightseeing pavilion, which provide residents a mutual interactive space.

Key Points of Landscape Phase Two

Frame shaping: different qualities of lines are applied to form regular collision and interweaving. The independent surfaces generated from the collision of lines fulfill and enrich the entire frame structure through the figure–ground relations of plants, floor and water.

Exquisite details: the application of modern decorative elements such as glass, metal, wood and stones in details is another view and experience the design wants to express to people. Through the collocation of these materials and exquisite sizes, a simple, concise but elegant landscape temperament has been forged.

Landscape experience on different levels: people oriented landscape concept directed architects to put residents' experience at the first place through enclosing of different materials, design of life pieces and creation of landscape atmosphere. Thoughts of space intended to using space as a medium to evoke residents of different ages, different background and different identities to think deeply about life and society.

FEATURE | 专题

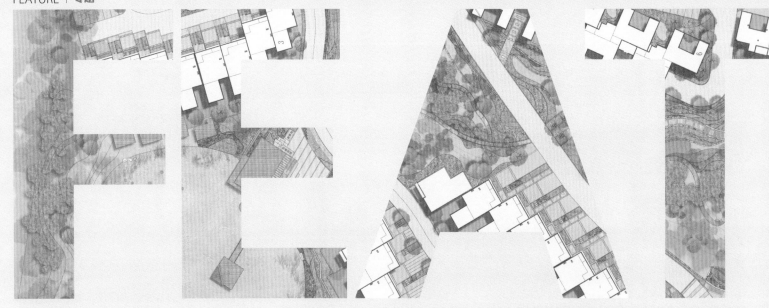

商业综合体

专题导语

随着商业地产的不断升温，以城市商业综合体为代表的新商业业态也异常火爆。作为一种复合型地产，其涵盖了办公、购物、娱乐、酒店、居住、文化等多种形态，内部各业态之间也形成了一种相互依存、相互助益的能动关系。商业综合体在建筑形态上往往表现为大众化、多功能、高效率的建筑群落。在设计上，商业综合体开始更多的考虑其选址、功能、交通、环境以及运营等方面的因素；同时，各种新材料和新技术的运用，也为高科技、高智能的商业综合体设计创造了良好的基础。本期专题将刊登知名设计师关于商业综合体的设计观点，探讨其设计要点，并精选优秀的项目实例与大家分享。

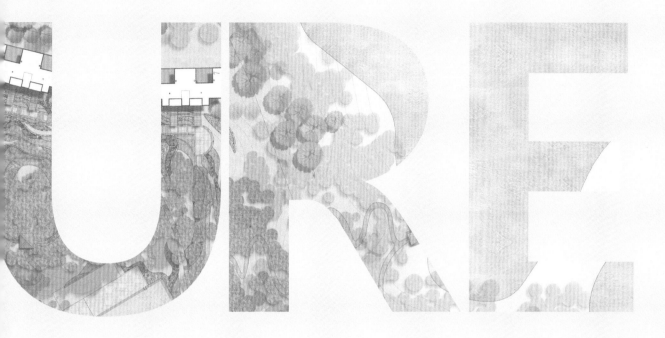

Introduction

With the heating of Commercial Real Estate, the new commercial activities oriented in urban commercial complex are extraordinarily hot. As a compound-type of real estate, it covers the office, shopping, entertainment, hotel, residence and cultural buildings, and develops a relationship of interdependence and mutual benefit among all kinds of commercial activities. Usually, commercial complex means the buildings group with the characteristics of popularization, multifunction and high efficiency. For design, it pays more attention to the factors of site selection, function, traffic, environment and operation. In the meanwhile, the application of new materials and technology creates good foundation for the high-tech and high-intelligent commercial complex. The current subject will introduce the well-known architects' opinion on the design of commercial complex and handpick excellent project cases to share with readers.

江苏新城地产
JIANGSU NEW TOWN REAL ESTATE

企业介绍

江苏新城地产股份有限公司创立于1993年，是以房地产开发与经营为主营业务的中国上市公司，公司具有房地产开发企业一级资质。公司名列中国房地产百强企业第20位，江苏省房地产业综合实力五十强第1位，2007中国华东房地产公司品牌价值TOP10第1位。

产品与服务

公司始终秉承"诚实做人，踏实做事"的企业精神，坚持"专业创造价值"的经营理念，凭借正确的市场定位和严格的内部管控，通过稳步扩张，逐步实现自身的发展壮大。公司业务范围目前涵盖常州、上海、南京、苏州、昆山、无锡等长三角核心区域。已成功开发规模化住宅小区50多个，为5万个家庭、15万业主提供了高品质的居住物业。

Company Profile

Jiangsu New Town Real Estate Limited established in 1993, which is a Chinese public company, specialized in real estate development and management, and possess qualification grade A of Real Estate Development Company. Ranked as No 20 in the best 100 Chinese real estate enterprise, and the best company in Jiangsu real estate industry, in 2007, it was ranked first in Top10 brand value real estate agency of eastern China.

Product and Service

Jiangsu New Town Real Estate Limited followed the enterprise spirit of "Be honest, Do best", persisting the operation philosophy of "Professionalism makes value", gradually achieve self-development and enlargement by means of accurate market positioning, strict interior control and steadily expanding. Their business area covered in Changzhou, shanghai, Nanjing, Suzhou, Kunshan, Wuxi and other core area in The Yangtze River Delta. Developed 50 scale residence community, provided 50,000 families and 150,000 owners with high-quality residential property.

广州合汇地产
GUANGZHOU HEHUI REAL ESTATE

企业介绍

广州市合汇房地产有限公司成立于2004年，是一家异军突起的新兴房地产有限公司。公司资历雄厚，人才济济，以广州增城市为主要发展地区，逐步向珠三角、江西、福建等一、二线城市拓展市场并不断发展壮大，土地储备已达4 000 000平方米；同时公司还拥有各类专业管理及技术人员。

产品与服务

广州市合汇房地产有限公司现正开发广州增城新塘汇东国际花园，项目临近107国道，位于新塘太平洋工业区，正在开发的汇东国际花园一期项目，占地9万平方米，建筑面积28万平方米，配套洋楼、小高层、高层建筑等多种建筑形态，建成后将为新塘"地王"式建筑群，依靠完备的配套设施，科学现代的物业管理，致力打造成为新塘顶尖商住楼。公司另有多个项目处于筹备中，致力于为广大的业主提供高品质、规范化的服务，打造合汇地产自己的精品房产品牌。

Company Profile

Guangzhou Hehui Real Estate was established in 2004, which is a burgeoning real estate limited company. With solid qualifications and great many of talents, considering Zengcheng as its main development area, gradually expanding to those first tier or second tier cities in Pearl River Delta, Jiangxi and Fujian etc, the land reservation has reached to 4,000,000 square meters, and full of various professional management talents and skillful talents.

Product and Service

Guangzhou Hehui Real Estate Limited Company are developing Xintang Huidong International Garden in Zengcheng District of Guangzhou, this project is located in Xintang Pacific Industrial Park and close to 107 National Road, the first phase of the this project covered an area of 90,000 square meters and floor area of 280,000 square meters, western-style building, small high-rise building and high-rise building form a complete set, it will become "prime site" architectural complex. Guangzhou Hehui Real Estate Limited Company devote to build top buiness living house in Xitang depend on their complete ancillary facility, scientific and modern property management. Besides, they have several preparing projects, which also devote to provide owner with high quality and normalized services, to make their own boutique real estate brand.

FEATURE | 专题

浅谈混合用途建筑的历史与未来

■ **人物简介**

罗曼·哈雷特
凯佳李建筑设计事务所伦敦总裁

罗曼·哈雷特 具有10年在波兰、美国、英国的从业经历，他将其在建筑、规划、室内设计和可持续发展方面的理念带到各个地方。罗曼·哈雷特拥有曾在大的合资公司（HOK）工作的经验，小的私有企业的从业经历（3xARC），同时兼有中等商业事务所的经历（FPA），最后于2008年建立起了其自己的公司（RHA）。他经验丰富，曾设计过阿联酋的大型发展综合体、商业写字楼，还有伦敦高端私人住宅项目。罗曼·哈雷特认为建筑必须同时具备漂亮、舒适和功能性。建筑需要满足其受众的需求，包括他们的文化和生活习惯。他通过分析现有的内容，当地的规章制度，受众的需要，理念和收益，个性化地对待每一个项目。最终的方案是一个审慎的设计过程和客户近似的合作。罗曼·哈雷特认为可持续设计可以改善我们的环境，降低能源需求、花销，并且提供一个健康的居住、工作和休闲环境。只要是具有实用性，他都试图用可再生或者低能源的技术运用到建筑中。

　　混合用途开发并不是一个新的概念，纵观人类历史，大多数的定居点都是基于混合使用环境这一概念建立的。在现代交通工具发明之前，大部分人的交通方式是步行，所以人们的生活和工作的地点常常在相同或相邻的建筑物里面。

　　随着时代的发展，科技的进步，在工业革命时期的英国，混合使用区域的面积开始减少。土地分区法的执行迫使工作和生活的地方分离。在那个时候,这一法规是非常有道理的，因为工业区排放的废气会影响附近居民的健康。工业革命也带来新的交通工具——铁路交通，人们因此每天到达更远的地方。所以，城市中心变成了商业区而城郊地区成为了居民生活区。但是，随着英国的去工业化，强制分区也变的不再那么必要。

　　在伦敦著名的圣保罗大教堂东侧，有一块被称为"一平方英里"的地方，这就是闻名世界的伦敦金融城。在这个区域里居住人口不足八千，但是在每一个工作日都会有超过三十万的人来到这里。可是一到周末，这个地方几乎成了无用之地，因为大部分的商店和餐厅都会关门。而伦敦的城郊生活相比之下就显得冷清了不少，那里的人口很少，空地却非常多。人们无论往返于工作场所、学校、或是购物中心，汽车（或者其他便利的交通工具）都是必不可少的。在这些年里，人们已经养成了一种非常低效的习惯。

　　在最近十年里，混合用途开发项目在英国比单一使用的项目得到了更多的投资，伦敦的许多地方都进行了翻修，从单一使用转换为混合使用。这些地方已经被视为商业，社会和美学上成功的典范。在金融城往东不远，有一个地方叫金丝雀码头，它是伦敦一个主要的商业区，但它曾经是一个矗立着数幢摩天大楼却异常冷清的地区。到了90年代，随着居住区，零售商店，酒吧和餐厅的不断进驻，现在这里变成了一个非常繁荣的街区。

　　从一个开发者的角度讲，通过混合使用开发，可以将风险分散到不同类型的建筑，而不会把所有的资本都投入到一个类型的建筑。因此，在地产市场中，如果写字楼销售不理想，住宅的销售可以为公司缓解财政上的压力。所以，在英国的商业地产中拥有住宅元素的往往会卖一个高价。

　　其实混合使用开发取得如此成功的道理很简单，是因为资源得到了最有效的配置。以一个高层为例，街面上的空间并不适合办公和居住，却是商业的黄金地点，而景色迷人的高层空间是公寓和酒店的不二之选。

　　关于能否在伦敦金融城里开发混合项目的争论持续了很长时间，一些人认为，在过去的25年里，金融城里面写字楼的价值没有明显的增长，所以适当的增加一些其他的使用形式可以吸引更多的投资，最大化利润。但是反对者则认为一旦办公空间被当做住宅以150年的契约卖掉的话，如果日后需要，它们将难以收回。

　　但在中国，混合使用开发的项目已经取得了巨大的成功，因为它在建筑密集度高的地区有着非常好的效果。一些多功能的建筑如

南京应天大街营房仓库商业开发

绿地花桥总部基地

聚源商业广场

■ 人物简介

李华
凯佳李建筑设计事务所设计师、市场总监
李华先生有不同的营销经验，行业范围包括：出版、电信和建筑。2010年，他在英国被授予特许市场营销人员，这使得他成为获此殊荣的最年轻人员之一。随后，他来到了中国，在天津大学学习，获得由中国政府提供的全额奖学金。李华有着非常丰富的项目经验，参与过包括住宅、酒店、博物馆、商业区和混合用途等多个项目。

集写字楼、酒店和购物中心为一体的大楼，或者拥有住宅区域的大型科研中心在中国如雨后春笋一般涌现出来。混合使用开发可以在城市景观内提高社会的整合度，带来更多的经济活动，商人们会更有信心的在这里租用店铺因为住宅区可以确保他们拥有稳定的客源，而商业区也会使他们的生活更加便捷。

在凯佳李建筑设计事务所，克雷·沃格尔带领众位杰出的设计师已经完成了中国许多的混合使用开发的项目，这其中包括了著名的长沙云龙数码科技园和沧州科技公园。尽管这些混合使用的项目都是有很多不同功能的建筑组成，但我们更强调的是整体的统筹规划而不是单一建筑的设计。设计混合使用的项目过程其实就是设计一个社会的过程。我们非常努力的给予长沙云龙数码科技园和沧州科技公园他们需要的空间，旨在保持生态平衡的同时鼓励人与人之间的交流和创新。

在各种不同的功能中，如何保持平衡显得尤为重要。然而，我们并没有一个完美的公式去参照，所以我们会更为注重人的需求。在中国，由于购物在人们的生活中的比重不断上涨，零售业会成为吸引人们到混合使用开发项目的关键因素。

开发混合用途项目的时候，其中的一大挑战就是人口密集度的计算，因为他决定了不同类型建筑的数量和他们将如何被使用。在大多数情况下，过高的人口密集程度会导致舒适度的下降和引发如水、电、交通等基础设施的分配问题。由于一个混合开发的项目的一大特性是其巨大的人流量，因此更为灵活的分配。在白天的时候，居民往往会去别的地方上班，而其他的地方的工人会来到这里工作，所以白天的时候商业会成为能源最大的消耗者，而随着夜幕降临，当商店关门之后，居民则更需要使用能源。

巨大的人流量往往将混合使用开发的项目变成更为安逸闲适的地方，从而吸引更大的人群。因为这里是一个可以居住、工作、购物和游玩的地方，加上设计师们的精雕细刻，必将成就一个与众不同，生机勃勃的传世经典。

Mixed-use Development: Past, Present and Future

Profile:

Roman Halat, RIBA, President of KaziaLiDesign, London office

With 10 years of professional experience in Poland, USA, and United Kingdom, Roman Halat brings his extensive knowledge of architecture, planning, interior design, and sustainable development. Roman has diversified his professional career by working for large corporate company (HOK), small private practice (3xARC), medium size commercial firm (FPA), eventually starting his own practice (RHA) in 2008. His experience varies from large mixed-use developments in United Arab Emirates, through office buildings, and high-end private residential projects in London. Roman thinks that architecture must be equally beautiful, comfortable, and functional. It needs to be customized for its final users, their needs, culture, and lifestyle. He approaches each project individually, with thorough analysis of existing context, local regulations, clients' requirements, ideas, and budget. The final solution is an outcome of careful design process and close cooperation with the client. Roman believes that sustainable design can improve our environment, lower energy use, bills, and provide a healthier living, working, and leisure atmosphere. Wherever practical, he tries to implement renewable or low energy technologies into buildings.

Mixed-use development is not a new concept, in fact throughout human history the majority of settlements have been based around mixed-use environments. Prior to modern transportation, most people travelled by foot, therefore people's living and work space were often in the same building or close by.

During the industrial revolution in the United Kingdom, mixed-use area's began to decline. Zoning laws were enforced to divide work and living spaces. At the time this made sense, since residential area's close to factories emitting pollution is harmful to public health. The industrial revolution also brought railway transportation, and so people were able to commute further distances on a daily bases than ever before. As a result, city centers evolved into business districts and the suburbs into residential areas. However, as the United Kingdom de-industrialized, the need to separate zones decreased.

The City of London, also known as "The Square Mile" is the financial district of London and has a residential population of less than 8,000 people. However, each day from Monday to Friday over 300,000 people commute to the square mile. At the weekend, this area is almost inactive, and most shops and restaurants are closed. London's sub-urban lifestyle is very isolated. The population density is very low and there is more space, however whether going to and from work, school, shopping centers or any other activity, a car (or some means of transportation) is required to get around. Over the years in the UK this has become a habit of sub-urban dwellers, but it is very inefficient.

In the last decade mixed use projects within the United Kingdom have seen more investment than single use buildings, and many areas of London have been restored, transforming from single usage to mixed use. These areas have been deemed as commercially, socially, aesthetically and environmentally successful. Canary Wharf, a major business district in London used be highly inactive and consisted of just high rise office buildings. Since the 1990's, residential, retail, bars and restaurants have since been added and the area is now a thriving district.

From a developers perspective, by building a mixed use area, they are spreading their risk across different building types, as opposed to investing all of their capital into one building type. Therefore, within the property market, if office space is not selling well, sales of residential buildings can cushion the finance. Furthermore, commercial property in the United Kingdom tends to sell at a high price where there is also an element of residential property.

Multi-purpose buildings make sense since different parts of a building are better suited to different functions. Street fronts to not suite office and living space, but work well for retail. However the upper floors, where the view is very good work very well for residential apartments and hotels.

Debates have been held whether to develop mixed use projects within "The Square Mile." Throughout the last 25 years the value of office space within "The Square Mile" has failed to grow significantly. Adding some form of mixed-use to these spaces could attract more wealth to the area. However, the other end of the debate is that if office spaces are to be converted into residential, and sold with leases of 150 years, then they would be difficult to turn the space back into an office if needed.

Within China, mixed use project have enjoyed huge success, mixed-use developments work very well in high density areas. Multi-function purpose

Profile:

Li Hua, Marketing Director of KaziaLiDesign

Ben has diverse marketing experience ranging in a number of industries including publishing, telecommunications and architecture. In 2010 he was awarded Chartered Marketer status, making him one of the youngest marketers to ever be awarded the accolade. He later moved to China when he was honored a full scholarship by the PRC government to study at Tianjin University. Ben has worked on a wide range of projects including residential, hotels, museums, retail and mixed-use.

buildings such has office towers and hotels with shopping malls attached to them, or large research centers with residential buildings have cropped up across the country. Mixed-use developments can increase more social integration and economic activity within the urban landscape. Retailers have more confidence in renting space in mixed use areas as they are almost guaranteed to have customers from the residential buildings, and from the residence perspective, the commercial buildings provide convenience.

At KaziaLi Design Collaborative Clay Vogel leads the design on mixed-use devlepoments and has completed several successful projects across China including the Changsha Yunglong Digital Technology City and Cangzhou Science and Technology Park . These mixed use projects, although may consist of a number of buildings with different functions, more emphasis needed to be given to the planning rather than the design on individual buildings. When designing a mixed use project you are also designing a community. We worked very hard to give the Changsha Yunglong Digital Technology City and Cangzhou Science and Technology Park the community space it needed so that it is both harmonious within the environment and encourages interaction and innovation between its inhabitants.

There also needs to be a good balance between the different functions such as residential, retail and office space, etc. While there is no perfect formula, the important factor is that people would want to live, work and play in the area. In China, shopping is an increasing leisure activity, and so retail can be a driving factor attracting people to a mixed-use development.

One of the challenges when creating a mixed-use project is calculating the density, this is due to the number of different building types and how they will be used. In most cases, a density too high will cause problems with amenity space, and infrastructure such as water, electricity and transportations, etc. However a mixed-use development with residential and commercial functions has a more dynamic flow of people and thus can be more flexible. During the day residence may commute to other places for work and workers from other area's will commute in. Therefore throughout the day most power consumption is taken up by the commercial are. During the evening, when shops and business close, and residence return home, there is a higher demand for power throughout the residential buildings.

This dynamic flow of people tends to make a mixed-use development a safer place and friendlier place, attracting people from the morning to night. They are a place to live, work, shop, visit, and play. The combination of bringing together two or more of the right functions supports each other to create a vibrant and successful place that is both distinctive and memorable.

拉斯维加斯商业街

天津宝利汉沽新开南路

上海文化金融城

关于商业综合体的设计关注点

■ **人物简介**

李颖悟
OAD欧安地建筑设计事务所总经理
国家一级注册建筑师
美国新泽西理工大学建筑学硕士
清华大学建筑学学士

Profile:

Eric Lee
General Manager of OAD Group
National first-class registered architect
M.Arch at New Jersey Institute of Technology
B.Arch at Tsinghua University

当今的城市商业综合体，已由过去从功能出发的单一模式向多元化方向发展。商业综合体作为涉及工作、居住、休闲等复合功能的建筑组群，对于解决中国城市的土地利用率，减少道路的交通压力以及城市的可持续发展等具有积极的作用。一个好的商业综合体项目，在准确的市场细分的定位条件下，需要更好的设计创意使所在的城市街区焕发新的活力。一种更人性化的生活方式在商业综合体中诞生发展，复合的功能聚集了高密度的人群，高密度人群带来蓬勃的商机，由此带来了土地价值的提升，品牌附加值的提高。成功的商业综合体将成为城市的新名片。

规划：城市肌理的延续而不是断裂。

商业综合体大多占据城市相对重要的区域，以往的模式经常是不同地段分属不同的发展商。发展商关心本地段的规划远大于与周边区域的关系，导致现今的公共空间缺乏、人行道路支离的城市隔裂状态。

商业综合体的复合功能决定它是一个开放的社区，一个成功的商业综合体就是一座缩微城市，在这里，可以实现各种各样的城市功能。而城市，因为高尚的生活质量而吸引人们集中，形成高密度的建筑和高密度的人群。因此，设计中应以人为本，按照人的尺度，尊重已有城市的肌理，借鉴那些有历史价值的古老城镇的空间组织，尽最大可能使可步行的公共空间在更大的范围延续，使人们在此能享受城市生活的乐趣，一种亲切的、人性化的、可以感知的空间尺度带来的舒适自在和优越感满足感。同时，妥善处理地面交通，整合地下停车，可以改善本区域的人车交叉的混乱和拥挤局面。

文化：面对未来，创造性的富有体验的场所制造。

应该开放地应用新工艺和可持续发展技术，突破固有的常规的建筑布局模式，探求每个项目与众不同的建筑和其周围环境相关联的个性场所特征。创造使人们具有愉悦的体验性的特色空间。

设计创新通常来源于我们每个项目的社会的、文化的、美学上的诸多因素，需对其进行深入研究；同时也来源于对建筑的使用者——"人"的关怀，敏锐观察分析人们的居住、工作和休闲方式，旨在强化人们生活方式的丰富性。作为个体存在的人，只有在群体中，才能找到自己的位置，人与人的互动，形成了各具特色的文化。作为传承文化的载体，建筑一直扮演着重要的角色。而群体复杂的行为方式，需要特定的空间与之配合，相得益彰。商业综合体的重点在于综合，富有特色的文化气质决定了综合体的前途，商业的成功依赖具有鲜明特色的文化基因的时尚表达和符合大众审美的生活方式与行为模式，这一切都来自于富有体验的场所制造。

设计中应重视建筑和环境的融合。强调城市公共空间、建筑和室内空间的整体性；尊重建筑所处的自然环境，环境景观使建筑充满生机和美感，淡化钢筋水泥森林的压抑，加强人们对自然生活的感知；尊重城市的历史，吸收传统文化，利用建筑空间表达出文化符号和场所精神；透过文化的视角审视今天城市生活，会发现，尊重传统就是面向未来。

建筑：尊重过去，表达今天。

人们的记忆里永远抹不去旧时对街坊、院落的浪漫感受。尊尚本土老建筑的永恒之美，在设计过程中应用当地材料，表达当地的建筑文化；同时，既需有面对未来、国际化建筑形式的丰富多姿，又需尊重过去，体现地方建筑的从古至今的文脉相承。建筑的语言可以丰富商业综合体的空间体验，加深场所感的制造。丰富体量的组合，多样性的材料的拼贴，对应的是五彩缤纷的城市生活。在汽车主导的城市中，大体量的建筑组合可以形成综合体的整体印象，而营造出的宜人尺度的内部空间，则需要恰当的细节设计和肌理组织。从整体到局部，从室内到室外，连贯紧凑的造型语言和空间组织，可以充分满足生活其中的人们的物质和精神的双重需求。

Concerns on the Design of Commercial Complex

Today's urban commercial complex has shifted its way from single function-oriented model to diversification. As a building group involved in work, housing, leisure and other complex functions, commercial complex plays a positive role in using land efficiently, reducing traffic pressure and promoting sustainable development of the city. Under the positioning condition of accurate market segmentation, a good commercial complex needs better creativity to revitalize the place where it stands. The birth of a more humane way of life in commercial complex gathers a crowd of high-density which brings along business opportunities, enhancing land value and brand value. Successful commercial complex will become the city's new business card.

Planning: Continuation of urban complex rather than fracture

Most of the commercial complexes occupy the relative importance in the urban area. The past pattern is different lots belong to different developers. Developers are concerned about the planning much more than the relationship with the surrounding area, leading to today's lacking public space and crack state of fragmented city sidewalk.

The composite function of commercial complex determines it is an open community. A successful commercial complex is a miniature city, and you can find all kinds of urban functions there, the gracious quality inside gathers the crowd, thus form a high-density building complex. Therefore, the design should be people-oriented, respect existing cities in accordance with the human scale, texture, drawing on the spatial organization of the historical value of the ancient town, and continuing the walkable public space on a larger scale as much as possible, so that people here can enjoy the fun of city life, a kind, humane comfort and superiority satisfaction. In addition, properly handle ground transportation and integrate underground parking can improve the chaotic and crowed situation across the area.

Culture: creating creative spaces of rich experiences in the future

We should start to use new technology and sustainable development technology, break the inherent conventional architectural layout mode to explore the unique building in each project and the characteristics associated with surrounding environment, creating featuring space that makes people happy.

Design innovation always depends on in-depth research on the social, cultural and aesthetic factors in each project and the observation on occupants' living, work and recreational mode. One can only find his place in groups, and the interaction among each people forms a distinctive culture. The architecture has always played an important role as a carrier of cultural heritage. Complex behavior in groups needs specific space to be linked with. Comprehensiveness is the key of commercial complex. Distinctive cultural temperament decides the future of the complex, commercial success depends on the fashion gene expression of the distinctive characteristics of the culture and the lifestyle and behavior consistent with the public aesthetic mode, all these come from creating creative spaces of rich experiences.

Attention should be paid to the integration of architecture and environment. We should emphasis on the integrity of the urban public space, architectural and interior space; respect the natural environment of the building, in which the building is full of vigor and beauty, and dilute the repression of reinforced concrete jungle to strengthen people's perception of the natural life; respect the history of the city, absorb traditional culture and make the use of building space to express cultural symbols and spirit of place; experience urban life through the perspective of culture, you will find that respecting tradition is facing future.

Architecture: respect the past, expressing the present

The romantic atmosphere in old neighborhood and courtyard is a lasting memory in people's mind. We should respect everlasting beauty of local old building and use local materials in design process to express the local architectural culture; at the same time, not only face colorful international architectural form in the future, but respect the past and reflect the context inheriting of local architecture. The architectural language can enrich the spatial experience of the commercial complex, deepening the creation of location sense. The rich combination of volume and the diversity of materials collage are corresponding to the colorful city life. In the car-dominated city, the combination of the general volume of the building can form the overall impression of the complex, but to create a pleasant scale of the internal space, you need the appropriate details of the design and texture organizations. From global to local, from indoor to outdoor, coherent and compact modeling language and spatial organization can fully meet the dual needs of the people's material and spiritual life.

FEATURE | 专题

浅谈商业综合体的设计及发展

■ 人物简介

陈晓宇
加拿大AIM国际设计集团董事长、总建筑师
加拿大注册建筑师
加拿大皇家建筑师会员
建筑学硕士

陈晓宇，华南理工大学建筑学专业，硕士。专注专业，二十年磨一剑。他不仅有二十年以上的建筑设计从业经验，更是国内首批研究MALL的先驱者，于1997年发表第一批关于商业地产研究的硕士论文。之后他在SHOPPING MALL的发源地——北美，有长达8年的探索与积累。在加拿大期间，先后就职于多伦多肯耐德建筑师楼（Kneider Architects），加拿大多伦多祖连积格建筑师楼（Julian Jacobs Architects），任主创设计师，并于2004年获得加拿大注册建筑师资格，成为加拿大皇家建筑师学会会员。随后于2006年在广州组建了以建筑设计为主，涵盖商业地产策划、景观设计、室内设计、广告策划等各专业无缝对接的设计产业链——加拿大AIM国际设计集团，并前瞻性地将商业地产作为集团研究的重点所在，在房地产行业遭遇冬天之际，以专业化、差异化展开突围之路，现集团已发展至200余人。由于其在商业地产领域独到的见解和建树，多次受邀各类商业地产嘉宾、评委。资深的行业背景、良好的方案水平、优质的设计服务团队，为AIM在商业地产领域积累了良好口碑。

商业综合体 "非理性繁荣"的背后。

商业综合体，从本质上来讲，是政府对城市用地高效、集约化开发的必然结果。商业综合体具有商业、办公、居住、酒店、展览、餐饮、会议、文娱和交通等几大功能，其基础结构是一种业态配比方式。符合绿色、环保的发展趋势。且随着经济的不断发展，国民收入不断拉高，大中城市人口密度不断增长，催生了城市对全天候购物环境的需求；与此同时，房地产行业经过了20多年的积累，众多发展商积累了足够的现金流、社会影响力，他们的发展需要通过打造商业综合体来实现可持续发展；且，住宅市场逐渐趋严的各项政策，使得大量资金亟待寻找出路；加上主营实业类的企业家，为了增强社会效益、提高企业知名度也逐步跻身商业地产的行列……不管从大环境、大趋势出发，还是城市经济发展、城市面貌改善、税收、就业等需要，抑或是企业发展、资金出路等方面来看，商业综合体，俨然成了"会下金蛋的鸡"，商业地产在多力的综合作用下，呈现出一派非理性繁荣。

群MALL乱舞 专业人员匮乏

长远来看，商业地产的前景是好的，它是城市化发展、升级换代的要求。我们从各方数据统计可以看到，新上马的和即将上马的MALL，仅佛山就30余个，广州也近30个，如此密集和数量庞大的商业综合体，如雨后春笋般的出现，群MALL乱舞的初步发展时期已经来临。可是，一个城市到底能承载多少商业综合体？这取决于城市的承受能力，如：城市的人口密度、城市的消费能力、城市的真正需求等。一哄而上、盲目上马，不考虑市场容量的非理性繁荣之后，将不可避免地出现阶段性过剩现象。

同时，由于国内MALL尚处于初步阶段，缺乏成熟经验，开发商集体转型的背后，还面临人才、专业知识等多方面难题，商业地产策划、设计、运营等全链条专业人员均十分匮乏，对商业和市场的了解不够，由此导致综合体在选址、定位、主题、设计、运营等方面较为粗放，且同质化严重，"千店一面"的过分复制屡见不鲜。城市综合体缺乏与文化的融合、商业空间的个性不足等弊端暴露无疑。

商业综合体的特性探寻

总的来说，商业综合体将会呈现出以下几个态势：

1 适合在亚洲发展。

商业综合体的发展离不开三个要素，经济增长、城市化、人口增长。随着亚洲经济的发展，世界商业综合体的主战场将会集中在亚洲。

2 依托中心交通枢纽。

交通与商业的关系自古便有诸多研究。由于交通枢纽带来的人流，物流，资金流，信息流产生的发散作用，使其极容易形成经济，商务，公共活动频繁的场所。而这将为商业的发展提供肥沃的生存土壤和广阔的发展空间。

如：华南地区最值得期待的宝兴永旺梦乐城。该综合体位于顺德容桂，总占地面积21万

宝兴永旺梦乐城 - 顺德容桂
(Baoxing Aeon Mall, Ronggui, Shunde)

金州体育城 - 贵州兴义
(Jinzhou Sports City, Xingyi, Guizhou)

金州体育城之"大峡谷梦乐城" - 贵州兴义
(Grand Canyon Dream City of Jinzhou Sports City)

m², 建筑面积约85万m²。该项目为典型的城市交通枢纽与商业相结合的城市商业综合体项目。轻轨、地铁、城市主干道都使该项目具备极佳先天优势。随着JUSCO的母公司永旺集团进驻该项目，使得该项目受到政府和行业的聚焦。在设计上，充分发挥交通优势，设计了"双首层"的主流线，大幅提升项目的土地使用价值和经济价值；为了尽可能降低项目风险，利于创造良好的投资环境，通过不同的业态组合对项目进行功能区分，有效避免由于体量过大而产生的同质化竞争；为了兼顾社会价值，在地块中心通过下穿处理，让出直径达40 m的中心广场，供市民休憩娱乐的同时，形成人流的聚散点，提升整个商业综合体的人气。

3 大型化发展。

大型城市商业综合体，能将城市中的商业、办公、居住、旅店、展览、餐饮、会议、文娱和交通等城市生活空间进行有机组合，并在各部间建立一种相互依存、相互助益的能动关系，从而形成一个多功能、高效率的综合体。城市综合体基本具备了现代城市的全部功能，其综合型、便利性、大而全的特点会使之成为一种趋势所在。

如：近期加拿大AIM国际设计集团与王志刚战略工作室共同打造的金州体育城。该项目位于黔西南锁钥之称的贵州兴义。整个项目占地3000亩（200万m²），建筑面积达370万m²，其中商业面积达100万m²以上。整个项目包含体育中心、学校、医院、会堂、博览馆、酒店、办公、商业、居住等功能。而其中最富有特色的是："大峡谷梦乐城"。该商业体不仅结合了当地景观——马岭河大峡谷，人文——5个少数民族居住区，地域环境——天蓝、地绿、水清，气候——亚热带湿润温和型气候等特征，而且建筑外部形象与内部经营相结合，塑造层层退台，将城市特征与商业特征很好的融合在一起，同时充分利用地形高差，在商业体内形成叠水景观，将景观、建筑、人融为一体。我们称之为：自然长成的城。

4 主题化专业化发展。

主题化综合体亦称体验式综合体，一般位于城市中心。该类综合体相对于大型化综合体来说，体量略小，针对特定目标人群，有明确的消费诉求和极强的经营特色。

如：极具岭南特色的"佛山瑞龙酒店综合体"。该项目是创意文化与传统文化相互补充、共生共荣的典范，创造了一种创新模式的文化商业理念。在研究当地的文化和市场后，确定了酒店、商业街、总部办公的业态组合。而"狮舞岭南、龙腾南海"的设计贯穿这个项目，卧龙、醒狮、蟠龙柱刻画出一个独一无二的文化主题的商业综合体。

对于一个商业体来说，提高其核心竞争力的关键主要有两个主要方向：大型化和主题化。大型化意味着能够提供一站式服务，其多样性选择的便利体验会倍受青睐；而主题化的商业综合体会通过层层定位明晰目标消费群体，相比于大型化的便利而言，主题化创造的是一种综合体验。而不管是哪个方向，精准的定位、良好的规划建筑设计和科学的运营管理都是决定其生死的关键所在。

商业综合体的设计呈专业化发展趋势

一个项目的成功，与其定位适合度、业态是否合理、能不能实现持续经营和可持续发展有很大关系，更离不开专业的设计和科学有效的运营管理。因此，策划定位、建筑设计和招商运营的同步介入和充分交流融合非常重要。

佛山瑞龙酒店综合体 - 佛山西樵
(Ruilong Hotel Complex, Xiqiao, Foshan)

如：即将开业的"番禺汇珑新天地"。该项目位于番禺市中心地带，建筑面积约14万m²。该项目在策划、设计、招商、推广等专业上的同步介入为该项目的成功奠定了基础。同时，火炬形式的主入口设计，充分考虑人流、车流的融汇，通过弧形的火炬，吸引来自多条道路上的消费者，使清河路与德新路交界处灼灼生辉。

由于商业综合体不同于住宅功能的单一，因此，在规划设计角度来讲，一个综合体在进行多种业态配比融合时，首先需要满足的便是不同业态对空间尺度综合平衡的要求。这也是商业综合体区别于住宅的难度所在。我们需充分考虑空间的可变性。比喻说，宴会厅对层高和面积的要求，在设计时候要充分考虑其柱网等方面与其他业态空间的连接问题。空间是否合理舒适，很大程度上考验设计师对空间尺度的把握能力。

如：位于东莞星河传说黄金位置的威斯顿酒店商业综合体，建筑面积近22万m²，引进LV、ZARA等一线品牌的高端定位，令万众期待。在多轮的设计中，由于充分考虑了空间的可变性，所以即使后期策划定位发生颠覆性变化、引入超五星级酒店、招商对象发生变化，由此导致的空间调整仍在可控的范围内。

整体来说，一个专业的设计团队在进行商业综合体设计时，需坚持以下原则：

1、以人的需求为第一要素
2、土地价值的最大化（含经济价值、社会价值、人文价值等）
3、与周边环境相融相生的原则
4、寻找项目的个性所在，量身定做的原则
5、人流动线合理，避免死角的原则
6、保持商业空间的灵活可变的原则
7、建筑外部形象与内部经营的一体化原则

在遵循这些原则的同时，还需重点考虑中庭、广场、停车场、出入流线、物流流线与运营成本控制等等。

番禺汇珑新天地 (Huilong Sunnyday Mall)

信德中心 (Shun Tak Center)

威斯顿酒店商业综合体 - 广东东莞
(Weston Hotel complex, Dongguan, Guangdong)

商业综合体设计的差异化

商业综合体设计中的差异化设计，不同于一般视觉上的"标新立异"，它需要充分结合不同地域的人口、人文、气候、消费习惯等因素量身打造，既要考虑城市形象，又要结合地块特征；既要充分体现经济价值，也要考虑社会价值、人文价值；既要充分考虑人的尺度舒适性，也要注重人的空间体验感。

如：信德中心。该项目位于南城国际商务中心地带，总占地4.5万m²，总建筑面积39.3万m²，地块限高80m。定位高端，可直飞澳门，分布有酒店、公寓、SHOPPING MALL、办公等业态功能。在设计上引入了"立体城市""都市信息站""全方位生活网络""绿色生态建筑"的理念，通过合理的入口设置，整合不同流线的空间关系等手法，打造了一个充满张力与活力的标志性建筑。充分体现了商业实用性和建筑美学的平衡。

商业综合体设计的环保

环保，是每个商业综合体都需要重点关注的问题，它不仅是全球设计趋势所在，也是对商业综合体持续运营的一大考验。谈及环保，必然谈及中庭及天棚。商业中心对光线的要求就像人需要呼吸一样自然。在设计中，我们既要考虑到光棚设计和项目定位的匹配性，也要充分考虑其在运营过程中产生的能耗。因此，我们在处理该类问题的时候，会通过一些科学原理，比喻"烟囱原理"的运用，最大程度实现自然采光的同时减少能耗。该类原理的运用，在"道滘客运站"项目就运用的很好，目前该项目白天不用开灯和空调，其通风采光都非常不错。

商业地产的大洗牌即将来临

外行扎堆、开发商集体进军商业地产，洗牌还有多远？一个商业地产的发展的黄金时代来临，伴随着阶段过剩，一个大洗牌时代也即将来临。如何衡量一个商业综合体的成功，我们不仅关注经济效益的最大化，我们更关注这个综合体是否也在社会价值、经济价值、城市功能、文化效益等方面实现了最大程度的融合。也惟有如此专业、专注的设计团队，才能在"群MALL乱舞"的时代，通过不断探寻差异化、创新与多元，实现可持续发展，树立行业口碑。

东莞道滘客运站 - 东莞
(Daojiao Station, Dongguan)

运用"烟囱原理"实现节能减排、良好通风采光
Applying "chimney principle" to save energy and get excellent ventilation and light

With Concise Remarks on the Design and Development of Commercial Complex

Profile:

Calvin X.Y. Chen
President and Chief Architect of AIM Group International
OAA
MRAIC
M.Arch.

Calvin X.Y. Chen, Master of Architecture in South China University of Technology, has more than twenty years experience in architectural design. As one of the pioneers in China to study on MALL, he published his master thesis on commercial properties in 1997. Afterwards, he came to the birthland of SHOPPING MALL - North America, staying and studying there for eight years. During his stay in Canada, he has ever worked in Kneider Architects and Julian Jacobs Architects as the principal architect. And in 2004, he got the qualification for OAA and became member of RAIC.

In 2006, he set up AIM Group International in Guangzhou which provides services on architectural design, commercial properties planning, landscape design, interior design and advertising planning. It focuses on commercial properties and develops fast even in the "winter of housing industry". Now it owns about 200 employees. With unique views and outstanding achievements in commercial properties design, Calvin X.Y. Chen has been invited as honored guest and judge for many activities. Years of experience, excellent planning skills, great design and service team provide AIM with a good reputation in commercial properties industry.

Commercial Complex: Behind the "Irrational Prosperity"

Basically speaking, commercial complex is the result of high-efficient and integrated development of urban land. Usually featuring the functions such as retail, office, residence, hotel, exhibition, restaurant, convention, culture and entertainment, transportation and so on, commercial complex is a structure composed by different programs which follows the trend of green and environmental protection. Moreover, with the development of economy and the increase of people's income, the population density in cities grows and it needs a 24-hour shopping environment. At the same time, with more than 20 years' development, the developers of real estate industry have obtained enough capitals and influence for developing commercial complexes and proving their strength. In addition, the housing market now faces tightening policies, thus the developers are seeking for new projects for investment. While big enterprises also join in this industry to increase social benefits and enhance their reputation. All these conditions make commercial complex development a best solution for many social issues. It is developed irrationally and prosperously with the supported of different parties.

Malls Everywhere: Lack of Professionals

In the long run, commercial property will have a promising future. It agrees with the development of cities. According to the relevant statistics, there are about 30 malls that are newly finished or will be finished soon in Foshan and Guangzhou separately. Shopping malls emerge like mushrooms after rain. However, how many commercial complexes can a city hold? It depends on the capacity of the city, such as the density of the population, the consumption ability and the practical demands. Without consideration to these conditions, irrational development will result in periodic excess.

At the same time, since it is at the primary stage of mall development, the developers lack of experience as well as professional knowledge and talents to develop, design and operate commercial complexes. Thus phenomenons such as simple reproduction and lacking of cultural connotations and characteristics appear in commercial complex development.

Features in the Development of Commercial Complex

Generally speaking, the development of commercial complex presents the following features:

1. Suitable to develop in Asia

Commercial complex cannot develop without three basic elements: economic growth, urbanization and increasing population. With the rapid development of Asian economy, the main battlefield for world commercial complexes will move to Asia.

2. Relying on Central Transportation Hubs

The research on the relationship between transportation and commerce can date from a long time ago. The transportation hub can attract people, goods, capitals and information for economic and business activities. It provides the basic conditions and development space for commercial complex.

Take the most expectative Baoxing Aeon Mall as an example. Located in Ronggui of Shunde City, this complex occupies a total land area of 210,000m^2, featuring a total floor area of 850,000m^2. It is a typical commercial complex that replies on transportation hub. Light rail, metro line and urban roads provide it with excellent convenience. It draws the attention of both the government and the public with the entering of JUSCO's parent company—AEON Group. In terms of architectural design, it takes advantages of the transportation convenience to set "double first floor" and increase the land value. To lower the risks and create a favorable investment environment, different programs are well arranged in different functional areas. In the center of the site, a sunken plaza with a diameter of 40m is designed for relaxation and entertainment, which also helps to attract people to the commercial complex.

3. Large Scale

Large-scale commercial complex can combine the spaces for retail, office, residence, hotel, exhibition, restaurant, convention, culture, entertainment and transportation together, making them rely on and benefit each other to form a multi-functional and high-efficient complex. Urban complex features almost all the functions of a modern city, and it characteristics such as integrity, convenience and large scale will become popular.

For example, AIM Group International and WZG Strategy Consultants have recently finished the project—Jinzhou Sports City, which is located in Xingxi —so-called the key to southwest Guizhou. Occupying a total land area of 2,000,000m^2, it has a total floor area of 3,700,000m^2, of which more than 1,000,000m^2 are used for commerce. The complex houses sports center, school, hospital, auditorium, exhibition hall, hotel, offices, retails and residences. The most distinctive program is the "Grand Canyon Dream City". It takes advantages of the local landscape resource—Maling River Canyon, local culture—five minorities, local environment—blue sky, green land, clear water, and local climate – mild and humid subtropical climate. Additionally, the form of the building comes with the internal programs to create setback floors. The elevation difference of the land is also used to form cascade landscape inside the complex. Thus it combines landscape, architecture and human beings together to create a naturally grown city.

4. Theme-based and Specialized

Theme-based complex is also called experience-based complex which is usually located in downtown area. This kind of complex has relatively small scale, providing characteristic products and services for specific customers.

Take Lingnan-style "Foshan Ruilong Hotel Complex" as an example. The project combines modern creative culture with traditional culture, and create an innovative cultural and commercial idea. After the analysis on local culture and market, it decides the combination of hotel, commercial street and headquarter offices. Local cultural elements such as the crouching dragon and dancing lion can be found in every corner of the commercial complex, creating a unique culture-based complex.

For a commercial complex, its competitiveness is decided by the scale and theme. Large scale means one-stop service, diversified choices and incomparable convenience. Whilst theme-based commercial complex has its target customers and it is dedicated to provide unique experience. No matter which direction it chooses, clear orientation, good planning and architectural design, as well as scientific operation and management will decide whether it will be a success or not.

The Design of Commercial Complex Becomes Professional

The success of a project is decided by its right orientation, reasonable programs combination, sustainable development, professional design as well as scientific operation and management. Thus it is important to do good communication on overall planning and architectural design.

Take "Panyu Huilong Sunnyday Mall" as an example. Located in central Panyu City, the project features a floor area about 140,000m². Planning, design, investment attraction and business promotion are considered at the same time, which will make the project a success. At the same time, torch-shaped main entrance is designed to benefit the circulation and attract more people here. It is the lightspot at the intersection of Qinghe Road and Dexin Road.

Difference from residential project, commercial complex will house multiple programs. Thus the design of it must first get the spatial balance between different functions. We should make the variability of spaces into consideration: for example, the banquet hall has a higher standard for height and size, thus the design must consider the column design and its connection with other functional spaces. Designers' ability to control the spatial scale will decide the comfortableness of the space.

For example, the Weston Hotel complex located in the gold area of Dongguan covers a total floor area of 220,000m², houses the world-class brands such as LV and ZARA, and will be a complex for luxury brands. During the planning and design, the variability of spaces is made into consideration which will meet the requirements for future adjustments and changes.

On the whole, when designing a commercial complex, a professional team must insist on the following principles:
1. People's demands first
2. Maximization of the land value(including economic value, social value and cultural value, etc.)
3. Keeping harmonious with the surrounding environment
4. Seeking for characteristics, custom-made principle
5. Reasonable circulation
6. Flexibility of the commercial spaces
7. Unification of architectural form and internal programs
Besides these principles, the team should also consider the atrium, plaza, parking, entrance and exit, logistic system, operation cost, etc.

The Design of Commercial Complex Features Differentiation

The differentiation in commercial complex design not means the unconventional in vision. It means that the design must consider the elements such as the population, culture, climate and consumption style of an area to create a unique impression and maximize the land value. It should also make people's comfort and space experience into consideration.

For example, Shun Tak Center, located in the CBD of Nancheng, occupies a total land area of 45,000m² and has a total floor area of 393,000m². The limit height of the buildings here is 80m. It is a high-end complex with hotel, apartment, shopping mall and office. With the ideas of "three-dimensional city", "information station", "one-stop life network" and "green eco architecture", we have designed a functional entrance and established a reasonable space system to create an energetic landmark building which shows the balance between functions and aesthetics.

The Design of Commercial Complex is Environment-friendly

All commercial complexes should pay attention to environmental protection. It is not only the global design trend but also the test for sustainable development. When talking about environmental protection, it will inevitably mention the atrium and canopy. Light to commercial center is just like air to human. In terms of design, we not only need to match the canopy with the project but also consider its energy consumption in the future. Thus, there is a solution by applying some scientific principle such as "chimney principle" to get the most daylight and reduce energy consumption. This principle is also used successfully in "Daojiao Station": it features excellent lighting and ventilation during daytime without using lights and air conditioners.

A Reshuffling of Commercial Properties is Coming

At present, laymen and developers stream in commercial properties market. When will the reshuffling come? The gold time for commercial properties accompanied with periodic excess is coming, thus the reshuffling will happen soon. How to judge the success of a commercial complex? We would not only focus on the maximization of economic benefit but also pay attention to the combination of its social value, economic value, urban functions and culture value. Only the professional and dedicated design team can realize the sustainable development of a commercial complex by seeking for differentiation, innovation and diversity.

TOP COMMERCIAL SPACE WITH SPECIAL FORM AND SEASIDE ELEMENTS | Bahrain City Centre

造型独特、蕴含海滨元素的顶级商业空间—— 巴林城市中心

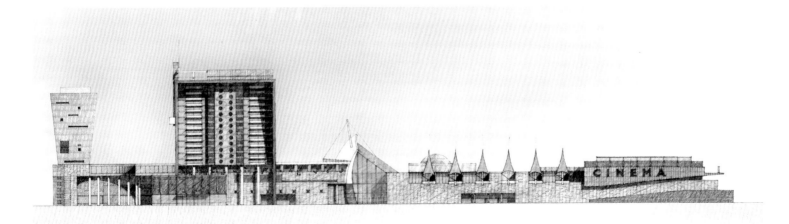

南立面图 South Elevation

FEATURE | 专题

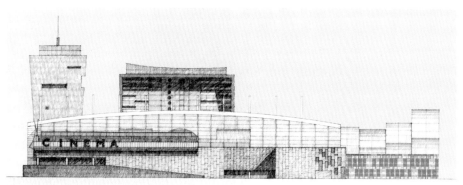

西立面图 West Elevation

东立面图 East Elevation

项目地点：巴林麦纳麦
建筑设计：RTKL
项目负责人：Ken Christian
建筑面积：150 000 m²
摄 影 师：Bill Lyons

Location: Manama
Design Company: RTKL
VP-in-Charge: Ken Christian
Size: 150,000 m²
Photographer: Bill Lyons

项目概况

巴林城市中心是繁华的麦纳麦地区的一个顶级的商业综合体。位于麦纳麦市中心的谢赫•哈利法•本•萨勒曼大道上，地理位置优越，拥有一个世界级的零售商业区，350个独家品牌和零售店；一个15 000 m²的室内/外水上公园；一

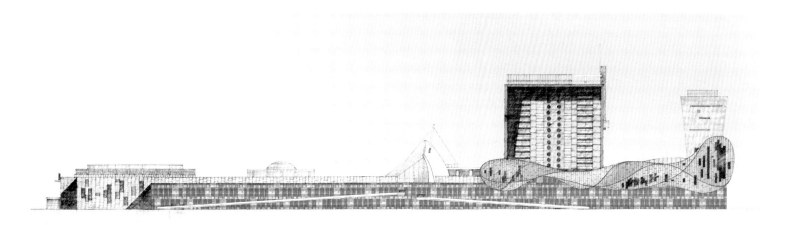

北立面图 North Elevation

个该地区最大的少年儿童娱乐中心；一个梦幻世界园；一个Cineco巨幕电影院；50多家 F＆B零售店和两家国际大酒店。总建筑面积超过15万m^2。RTKL的国际设计水准，使巴林购物中心近来在国内外广受好评。它是中东地区的首个世界级商业综合体项目。满足了这个人口稠密地区的需求。

规划布局及建筑设计

签名式的商场标识呈动态波浪状的结构，购物中心楼体是该项目的主体部分。中央广场是购物中心多层次空间的汇聚地，不同的色彩、灯光和造型，让整个大楼的每一处都栩栩生辉。

设计灵感来自海滨图像——冷色调，郁郁葱葱的风景和柔和的光线的内部空间与周围炽热和稠密的环境对比让人耳目一新。在凉爽的傍晚，室外露台和风景优美的花园，再加上高档餐厅，给人们提供了亲密的场所。

游客可以享受4或5星级酒店的水疗和屋顶游泳池服务，然后再去中心广场，那有移动水上乐园、多屏幕电影院和许多餐馆；同时还有家庭娱乐中心、美食广场和水上乐园。水上乐园一年四季都可以在室内和室外活动；此外还有配套的餐厅和多功能厅等设施，让所有人都能乐在其中。

FEATURE | 专题

Profile

Bahrain City Centre is a trend setting, mixed-use development designed to serve as an oasis from the heat and bustle of Manama. Strategically located on Sheikh Khalifa Bin Salman Highway in the heart of Manama, Bahrain City Centre offers a world-class retail experience with 350 exclusive brands and retailers spread over 150,000 square metres; a 15,000 square metre indoor/outdoor water park, the largest in the region; a family entertainment centre for children and young adults, Magic Planet; a 20-screen Cineco Cinema; over 50 F&B outlets and two international hotels.

Recently opened to national and international acclaim, the development showcases RTKL's expertise in delivering world class multi-disciplinary design services throughout the Middle East. It is the first of its kind in this location and meets the

demands of a sophisticated regional demographic. .

Planning and Design

The signature attraction is a dynamic wave-like structure that serves as the project's backbone. The multi-level space provides a central plaza that facilitates easy circulation throughout the various floors and functions as an exciting gathering place. A mix of colour, lighting, and dramatic forms not only highlight the centre's individual components, but also knit the entire development together.

The architectural language is derived from waterfront imagery—cool colours, lush landscaping and filtered light make the internal spaces a refreshing contrast from the heat and intensity of the external environment. In the cooler evening hours, outdoor terraces and landscaped gardens are coupled with high quality restaurants to create intimate nodes of activity around the project.

Visitors will be able to stay in either the 4 or 5 star hotels with spas and rooftop pools before exploring the Central Plaza, move on to the Waterworld or perhaps enjoy the multi screen cinemas and many restaurants. The project has an extensive family entertainment centre which is linked to the food court and Waterworld experience. The Waterworld spaces are designed for use throughout the year with both indoor and outdoor activities. The pool facilities are supplemented with restaurants and function rooms to ensure that the experience suits the whole family whilst being capable of accommodating other social functions.

HORIZONTALLY DISSECTED SKIN AND ENCLOSED LAYOUT | De Kameleon

水平切割的表皮　围合形式的布局——Kameleon商住综合体

项目地点：荷兰阿姆斯特丹Bijlmermeer
客　　户：Principaal / De Key
建筑设计：荷兰NL建筑师事务所
项目团队：Iwan Hameleers、Gertjan Machiels；Pieter Bannenberg、Walter van Dijk、Kamiel Klaasse；Barbara Luns、Gen Yamamoto、Ana Lagoa Pereira Gomez、Jouke Sieswerda、David de Bruijn、Jung-Wha Cho、Florent Le Corre、Stephan Schülecke、Tomas Amtmann、Joao Viera Costa、Jorge Redondo、Juerg-Ueli Burger、Nora Aursand Iversen、Kim Guldmand Ewers
总实用楼面空间：55 500 m²
地块面积：9 250 m²
摄　　影：Luuk Kramer、Marcel van der Burg

Location: Bijlmermeer, Amsterdam, the Netherlands
Client: Principaal / De Key
Architectural Design: NL Architects
Project Team: Iwan Hameleers, Gertjan Machiels (Project Architects); Pieter Bannenberg, Walter van Dijk, Kamiel Klaasse(Design); Barbara Luns, Gen Yamamoto, Ana Lagoa Pereira Gomez, Jouke Sieswerda, David de Bruijn, Jung-Wha Cho, Florent Le Corre, Stephan Schülecke, Tomas Amtmann, Joao Viera Costa, Jorge Redondo, Juerg-Ueli Burger, Nora Aursand Iversen, Kim Guldmand Ewers
Gross Useable Floor Space: 55,500 m²
Plot Size: 9,250 m²
Photography: Luuk Kramer and Marcel van der Burg

项目概况

位于荷兰Bijlmermeer的Kameleon住宅小区是一个超大型的公寓建筑群，里面设有购物中心、街心公园及大量停车位。Bijlmermeer地区是一个犹太人集聚的地方，由于受当时的激进主义影响，尽力的想把其变成一个荷兰传统的城镇风格。小区内的一栋十层高的建筑，在当地很有异域风情。

规划布局及建筑设计

Bijlmermeer地区的铁路交通非常有特色，而住宅区就建在Karspeldreef沿线上，在看上去如此狭小的地方建立这样一栋建筑，实在是令人叹为观止。建筑底层是新的购物中心，与公共活动区域相连接。正面看上去，造型有点像数字8。这样奇特的造型非常有利于经济发展和便利交通。

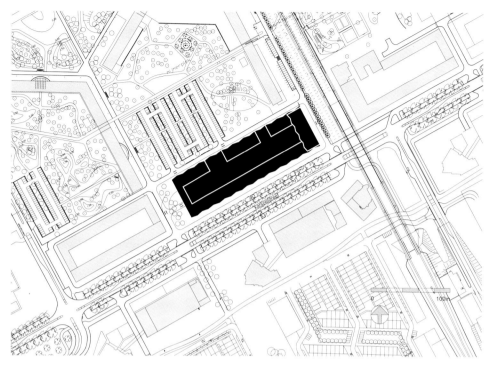

总平面图 Site Plan

FEATURE | 专题

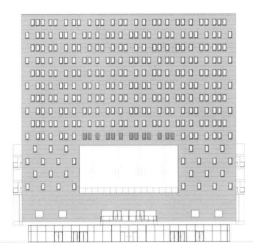

50m

东西立面 East and West Facade

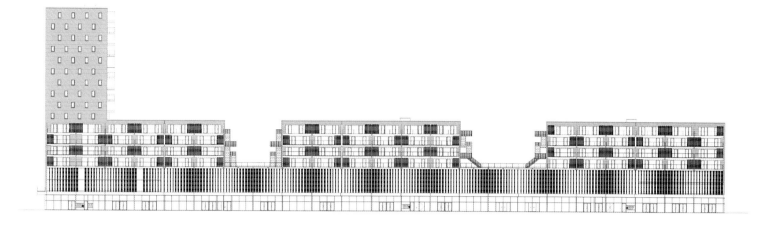

50m

北立面 North Facade

50m

南立面 South Facade

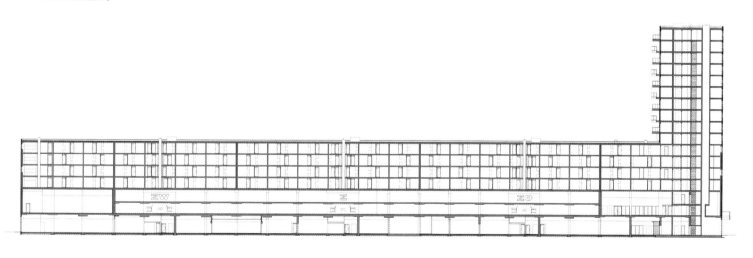

50m

纵切面图 Long Section

NEW HOUSE_083

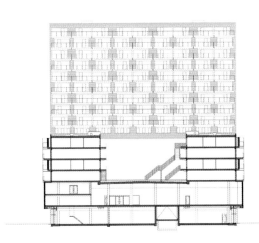
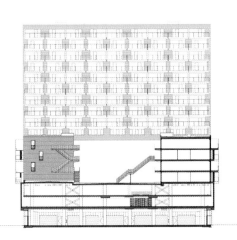

50m

横切面图 Cross Sections

在购物中心上方建立停车场实在是物超所值。停车场的一端通往超市,另一端则通往餐饮区及健身中心。所有住户、进入商场、超市的人们都可以将车停在上面,十分方便。停车场向两侧敞开,以获取足够的自然光照。在停车场上方有一个非常大的花园,内有十二棵大树及一条河流。

小区内的公园是由四层高的、拥有168套单位的公寓楼围合而成,既保证了公园的清静,又提供了很好的休闲场所。居民可以在公园里烧烤、运动、玩耍。建筑的外部表皮被水平的切割和不同的表皮处理方式打乱,从而减少了垂直方向上的压抑感。一个八乘八米的方格形的布局减少了结构的建造成本,同时生成一种富有节奏感的Z字形的阳台布局,有效地保护了住户的隐私。一个巨大的空洞位于坚实的立面之中,它充当窗口,人们因此可以看见室外的花园。

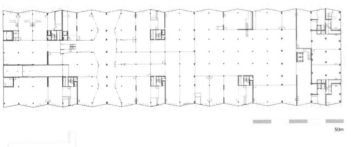

地面层平面图 Ground Floor Plan

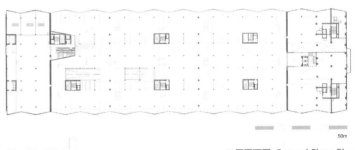

二层平面图 Second Floor Plan

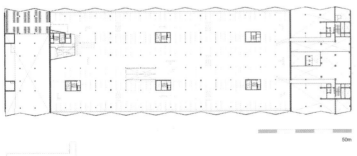

三层平面图 Third Floor Plan

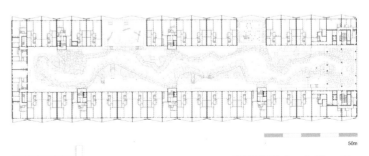

四层平面图 Fourth Floor Plan

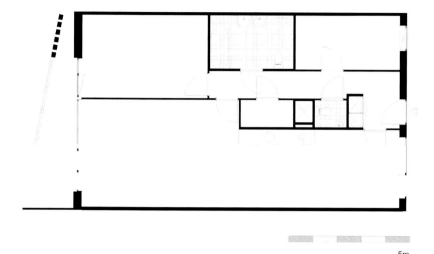
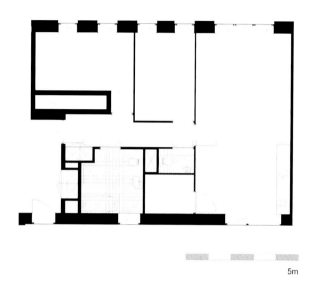

5m

5m

C1 户型 Housing Type C1

D3 户型 Housing Type D3

FEATURE | 专题

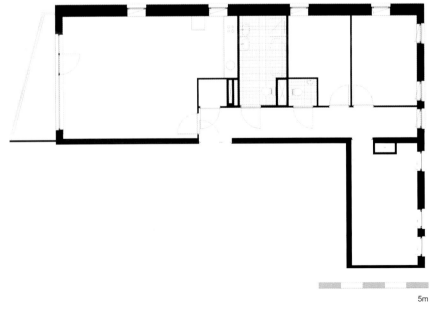

T6 户型 Housing Type T6

Profile

De Kameleon is a supersized housing block including a new shopping center and plenty of parking in the area formerly known as Bijlmermeer.The Bijlmer is the one area in the Netherlands that sometimes is considered a ghetto. At the moment the area is going through a radical renovation process: an attempt is being made to turn it into a regular Dutch suburb. Standard low-rise housing is introduced that replaces the 10 storey apartment buildings but also the green spaces in-between them. In spite of the new format, the Bijlmer remains exotic: it is the place to be for a sensational Roti or sundried Bats.

Planning and Architectural Design

The Bijlmer features a fantastic elevated subway track, maybe the only suitable backdrop for an R&B video in the Netherlands. De Kameleon is placed along the Karspeldreef, one of the main arteries in the area. It is quite a surprise that amidst the new ideology of the small scale such a large new building is projected. Kameleon is organized in horizontal slices. On ground floor is the new shopping center. All shops are accessible directly from public space, there is no collective interior: Kameleon is not a Mall.

The supermarket, normally a bulky program with extensive impenetrable facades is embedded in smaller units that as such both differentiate and activate the 'plinth'. There is one shortcut, The Passage, at 2/3rd of the length, creating an '8'. The 8 is good for circulation and good for business. From here an escalator connects to the next level, continuing the 8 in the 3rd Dimension. On the 2nd floor is one more supermarket; easily accessible from the public parking on the same level.

Positioning the parking on top of the shops proofed to be cheaper than in a basement. The parking is 'charged' by the supermarket on one end and food court / fitness center on the other. Since these programs feature large floor to ceiling heights an extra parking level fits in. The residents will park their cars here. The facade of the parking is open to the sides allowing natural ventilation. A very large garden is placed on top of the parking. It includes 12 serious trees and a river!

The garden is surrounded by a four story housing block containing 168 apartments. The side facing the Karspeldreef is continuous to protect the garden from street noise and to create an 'urban wall'; the other side facing the typical hexagonal green space is punctured. The gaps can be used as playgrounds and for BBQ's. The repetitive structure makes the project affordable. The rhythmic building bays of 8 meter and the parking and shopping grids correspond nicely. Every other carrying wall is extended to support the balconies and to provide privacy. The large balconies create dynamic patterns. Winding stairs lead to the garden and differentiate the large courtyard. A 10 story slab with 58 apartments rests on this flat Block. It creates a counterpoint to the horizontality and becomes a 'billboard' facing the subway. A Supersized window visually connects the elevated subway and the elevated garden that are precisely the same height.

OPEN AND SCENIC URBAN COMPLEX BUILT NEARBY THE MOUNTAIN

| Weiyi International, Zunyi, Guizhou

开放立体、依山而建的景区化城市综合体—— 贵州遵义唯一国际

项目地点：中国贵州省遵义市
开 发 商：遵义浙商房地产开发有限公司
建筑设计：上海中建建筑设计院有限公司
总用地面积：95 811.84 m²
总建筑面积：667 690 m²
容 积 率：5.5
绿 地 率：30%

Location: Zunyi, Guizhou, China
Developer: Zunyi Zheshang Real Estate Development Co., Ltd.
Architectural Design: Shanghai Zhongjian Architectural Design Institute Co., Ltd.
Total Land Area: 95,811.84 m²
Total Floor Area: 667,690 m²
Plot Ratio: 5.5
Greening Ratio: 30%

项目概况

项目位于贵州省遵义市汇川区，昆明路东西两侧、大连路南、香港路北。整个项目集商业、景观、人文、居住于一体，将打造成遵义中心城区最大的国际水平景区化城市综合体。包括建成贵州省首个国家4A级商业旅游景区、贵州省唯一景区化RBD，也是全国首个立体全景式时尚商业中心；创建遵义最时尚最富特色的城市新地标，引领都市时尚快乐新生活，缔造黔北"小香港"，成为贵州省首个回乡创业"小老板计划"基地。

规划布局

该项目属于典型山地建筑，总体布局结合地形依山而建，在基地内沿昆明路两侧布置主要沿街商业，中间设置一大型超市，南北各设一个商业广场，满足商业活动，人流疏散以及停车所需场地；昆明路南侧沿街设计两栋精品酒店式公寓，北侧设计一栋200 m高地标性超高层办公酒店综合楼；基地北侧设置8栋高层住宅楼。北高南低，围合布局，在小区内部形成完整的集中绿化景观，提高住区生活品质。

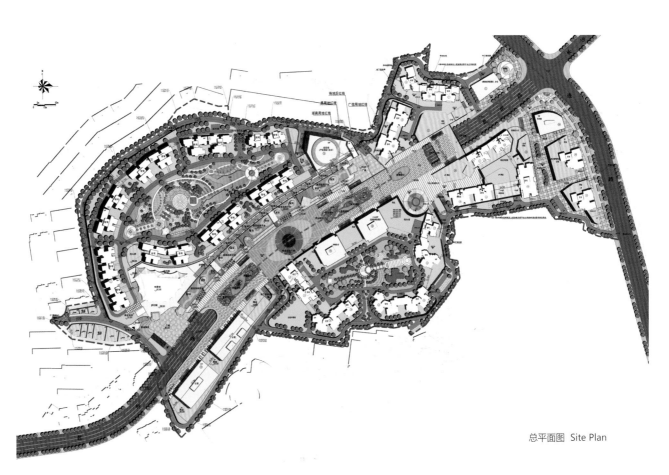

总平面图 Site Plan

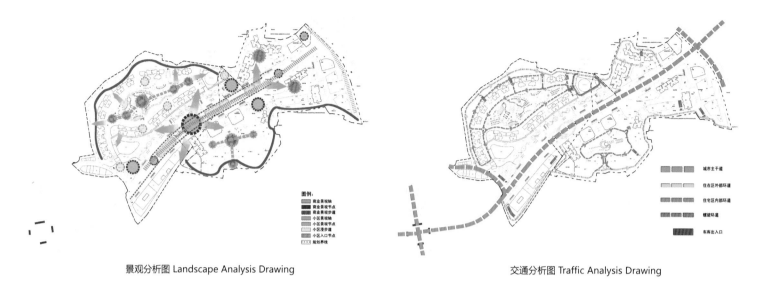

景观分析图 Landscape Analysis Drawing   交通分析图 Traffic Analysis Drawing

建筑设计

建筑结合地形设计是本案最重要特色，在典型的山地城市中处理好建筑与地形的契合；双首层、天街、商业平台广场等概念在设计中一一展现；大，中，小混合型商业共同营造出一个布局合理、功能齐备、交通便捷、绿意盎然、生活方便，具有文化内涵的商住氛围。

项目打造的三维立体情景式休闲商业街，采用两个商业广场作为开放节点引入大量人气，同时用内街增加了商业内部界面，激活了原来并不沿路的商业店铺。形成品牌，提升了地块的商业价值，同时在入口的两个商业广场中引入了地上一层和二层步行街的概念，让人的活动由自动扶梯电梯等引入地上一层和二层的空中步行街，激活了更多的商业店铺界面，最大化的利用了商业空间。天街层建筑以欧美建筑风格为特色，辅以西式星座景观雕塑以情景式商业为模式，引领都市时尚生活方式，打造遵义城市最具影响力和美誉度的新地标。

景观设计

在景观上，商业部分层层跌落、互为风景；本设计在强调商业价值的同时，增加了一种附加价值就是顶部风情商业天街，形成了良好商业街区氛围和商业景观带。根据规划要求，基地内绿地率设计为30％，满足规划要求；绿化布置主要为基地周边绿化带以及建筑庭园广场内的绿化布置；同时为了营造良好的视觉环境，在商业裙房的屋面也设计了屋顶绿化。

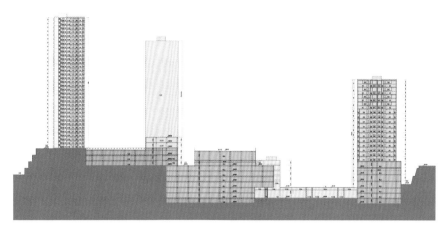

1-1 剖面图 1-1 Section

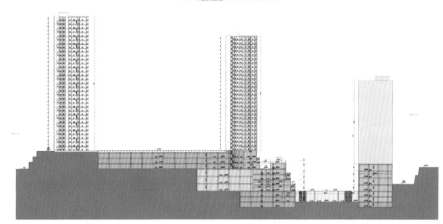

2-2 剖面图 2-2 Section

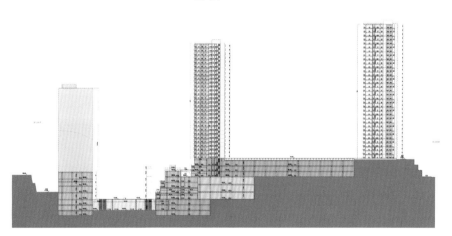

3-3 剖面图 3-3 Section

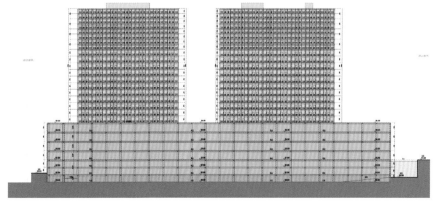

5-5 剖面图 5-5 Section

FEATURE | 专题

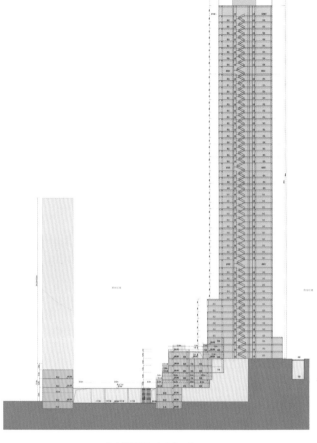

4-4 剖面图 4-4 Section

Profile

The project is located in Huichuan District of Zunyi City, developed along Kunming Road and sitting on the south of Dalian Road and the north of Hong Kong Road. Integrating retail, landscape, culture and residence, it is envisioned to be the biggest international scenic urban complex of downtown Zunyi. It will include the first national 4A-grade commercial and tourism area, which will be the only RBD in Guizhou province and the first three-dimensional landscape commercial center in China. As the new landmark for Zunyi which leads to modern happy life, the project will be "small Hong Kong" in north Guizhou Province and the first base for small business.

Planning and Layout

As the typical mountain buildings, the overall planning is made according to the topography of the mountain. Within the site, retails are arranged along Kunming Road, a supermarket is located in the center, and two plazas are set on the south and north end for commercial activities, circulation and parking. On the south of Kunming Road, two boutique hotel-style apartment buildings are built; on the north, it is a 200m high landmark building for office and hotel. Other eight high-rise residential buildings are set in the north of the site. Buildings are higher in the north and lower in the south to provide perfect layout and excellent green landscape which leads to a high-quality lifestyle.

Architectural Design

Buildings of this project are designed according to the topography: double first floors, sky streets, and platform plazas are designed; mixed retails of different sizes are combined together to create a green cultural neighborhood with reasonable layout, multiple functions, as well as convenient traffic and living conditions.

The three-dimensional commercial street features two commercial plazas to attract people and provide more frontage surfaces for retails. The first and second floor of the plazas are used as sky pedestrian streets to maximize the commercial value. The buildings along sky streets are designed in European-American style with constellation sculptures to be the most influential landmark in Zunyi City.

Landscape Design

The commercial floors are landscapes for each other. Commercial street is also set on the top floor to create good commercial atmosphere as well as additional landscape space. According to requirement of the planning, the greening ratio of the site reaches 30% with green belts mainly the periphery and internal courtyards. Meantime, for visional consideration, the commercial podiums are designed with roof greens.

FEATURE | 专题

ORGANICALLY COMBINED ARCHITECTURAL FORM CONTRACTED AND DYNAMIC URBAN SPACE

| Changsha Shimao Plaza

有机组合的建筑形态 简约动感的城市空间
—— 长沙世茂广场

项目地点：中国湖南省长沙市
开 发 商：长沙世茂投资有限公司
建筑设计：上海中建建筑设计院有限公司
合作建筑设计：美国捷得国际建筑师事务所
总用地面积：16 073.14 m²
总建筑面积：229 156.68 m²
容 积 率：12.30
绿 地 率：25%

Location: Changsha, Hunan, China
Developer: Changsha Shimao Investment Co., Ltd.
Architectural Design: Shanghai Zhongjian Architectural Design Institute Co., Ltd.
Cooperative Design Company: The Jerde Partnership
Total Land Area: 16,073.14 m²
Total Floor Area: 229,156.68 m²
Plot Ratio: 12.3%
Greening Ratio: 25%

项目概况

项目位于长沙芙蓉区肇家坪，临建湘南路，向北建湘南路与五一大道相交，沿五一大道向东直通长沙火车站、向西通向湘江一桥，沿建湘南路向南直达芙蓉中路。东侧有凤凰街，南侧有潘后街，东南侧为老城区建筑。基地临近芙蓉广场，且烈士公园、湖南省博物馆、贺龙体育休闲娱乐广场等遍布周围，地势较平坦。基地总用地面积为16 073.14 m²，净用地为13 827.34 m²，地上为75层，地下为4层。周边生活配套设施完善，交通便利，地理位置优越，极具发展潜力。本案力图创造一个富有活力的、具有国际化理念、现代时尚的标志性建筑。

规划布局

建筑由商业裙楼和办公塔楼两部分组成。整个建筑形体成一柔美的曲线弧形，与旁边地块的裙楼很好的对接上。商业主入口位于东北侧，入口位置规划为广阔的入口广场，使其容纳性更大、视野更广阔，并形成了很好的景观视野。办公的主入口位于西侧及东北侧，有较大的绿化空间及广场空间，使其入口更加开阔并不失生态。在东侧还有餐饮单独的出入口，使其功能更加合理。交通流线组织亦清晰明了，布局合理。

建筑设计

整个建筑简洁、大气、现代、个性鲜明，极具标志性。本方案在形体创作时强调了城市的外部空间环境和脉络设计，整个建筑轮廓线丰富，建筑空间形态组合关系有机协调，建筑有机的体

总平面图 Site Plan

量组合及其光影变化具有强烈的节奏感和韵律感，建筑风格端庄大方、高雅脱俗又清朗简洁、新颖夺目，创造了完整的形体和气势。

项目受到传统中国画特性启发，以"峡谷"和"高山瀑布"为设计理念，立面设计多采用自然符号。基地沿建湘路的主要沿街面，一座5层高充满扩张力的弧形商业裙房有力的稳固在A地块，使空间形成一个开阔圆弧形广场，视野开阔。

商业裙楼立面富有变化，商业气息浓厚，弯曲的弧形流线使建筑更具层次感，犹如层层叠叠的山峰，给人以气势磅礴之感。塔楼的立面设计采用玻璃幕的"高山瀑布"，在玻璃和LED照明反射下融合出一种水往下流淌的动感，汇聚在中央广场中。顺着水流而下，瀑布浇注在每一层楼中，形成了层层的空中花园。

景观设计

景观设计包括：道路绿化、商业广场、景观小品。道路绿化以常绿行道树装点，人行道铺地以及广场砖铺贴。商业广场采用硬质铺装与景观花池绿化相结合的手法，增强空间的活跃感。同时布置活动场地、健身场地、艺术廊等。

沿允嘉巷立面　　沿建湘南路立面　　沿规划道路立面　　沿东庆街立面

Profile

The project is located in Zhaojiaping of Lotus District in Changsha, facing Jianxiang South Road and intersecting with Wuyi Road. Wuyi Road leads to Changsha Railway Station in the east and Xiangjiang River First Bridge in the west while Jianxiang South Road leads all the way to Lotus Middle Street in the south. The project neighbors Phoenix Street on the east, Panhou Street on the south and old district architectures on southeast. The base is also near to Lotus Square, martyr memorial park, Hunan Provincial Museum and Helong Sports and Leisure Plaza, etc. The entire base covers an area of 16,073.14 m^2 and floor area of 13,827.34 m^2 with 75 floors over ground and 4 floors under ground. The site enjoys superior geographic location and development potential with complete living facilities and convenience transportations in the surroundings. The project will be forged into a dynamic, international, modern and fashionable landmark building.

Planning and Layout

The project consists of commercial skirt building and office tower, which takes up a graceful curve connecting rightly with the skirt buildings next to the site. The main entrance is located in northeast direction, where is planned to be a spacious square to accommodate more and provide larger vision and views. The main entrance of office building is located in the west and northeast direction, which enjoys large green space and square space to effectively broaden the entrance area and increase ecological part. In the east direction, there sets up an independent entrance for catering to fulfill the function of the project. The circulation of the project is clear and well organized.

Architectural Design

The entire space is simple, generous, modern and characteristic. In design process, this project gives much attention to the outside environment of the city and contextual design. The entire building possesses abundant architectural contour lines, coordinated architectural syntagmatic relation, organic mass combination and strong rhythm and tempo in light and shadow changing. The architectural style is elegant, dignified, clear, bright and attractive, creating a complete body and atmosphere.

The project is enlightened by Chinese traditional paintings. With valley and high mountain waterfall as the main concepts, the elevation applies natural elements on the projects. On the side along Jianxiang Road, a five-floor high commercial skirt building with expansionary force is fixed on Site A, which forms a spacious circular arc with wide horizon.

The skirt buildings convey strong and intense commercial atmosphere and the bending arc adds stronger sense of levels, which is reminiscent of the mountains upon mountains, grand and momentous. The elevation design for tower building adopts "high waterfalls" of glass curtain to form innervation of water running down with the reflection of glass and LED lights and gather in the middle of the square. Following the water down, the waterfalls pour on each level to give birth of gardens in the air on different levels.

Landscape Design

Landscape design include road greening, commercial plaza and sketches. Road greening is carried out with evergreen trees, sidewalk and plaza stones. The commercial plaza applies hard pavement in combination with framed flower bed to increase the dynamism of space. Meanwhile, activity ground, exercising ground and art gallery are also equipped in the place.

CHANGEABLE URBAN COMPLEX FULL OF ORDERLINESS AND LOGICAL SENSE | Zhangjiagang Shopping Park

充满秩序与逻辑感的多变城市综合体—— 张家港休闲购物公园

项目地点：中国江苏省张家港市	**Location:** Zhangjiagang, Jiangsu, China
开 发 商：张家港城市投资发展有限公司	**Developer:** Zhangjiagang City Investment Co., Ltd
建筑设计：马达思班	**Architectural Design:** MADA s.p.am.
占地面积：33 943 m²	**Land Area:** 33,943 m²
建筑面积：194 594 m²	**Floor Area:** 194,594 m²

项目概况

张家港购物休闲公园处于张家港市城西新区，位于张家港城西新区中心地带，国泰路以西，距离市中心2 km，项目用地在整个购物公园基地内的最北侧，北临大寨河，南为整个购物公园基地。它既是一个风景优美的公园，又是一个集购物、餐饮、休闲、娱乐、文化、居住、办公于一体的大型商业综合中心。在商业、办公、居住、休闲、环境等生活的多方面提出了自身独特的新颖理念。

规划布局及建筑设计

整个项目设计采用点、线、面相结合的方式，在庭院内有水池、绿化、休闲座椅，形成繁华优美的购物环境，室外将广植灌木、花卉和草坪，通过色彩与季节的搭配，形成四季丰富多彩的购物娱乐环境。

欧风街由十个不规则单体组成，外围墙面厚重封闭，呈古堡风貌，具有保护性质，内部由单体围合构成三个内向庭院，庭院内有水池、绿化、休闲座椅，形成繁华优美的购物环境，连接信息管理中心、文化中心和公园山丘的空中走廊贯穿其中。整个欧陆风情街性格内省，外表朴素大方，内在繁华热闹，风情万千。朝向内庭院一侧平面上不规则的部分被整齐地切割出来，构成中庭空间，此处的立面由细巧的木框幕墙构成，与外围的砖墙立面形成对比。屋顶采用多折的坡形屋面，具有中国传统民居风格。除了平面的不规则外，立面及屋顶也呈不规则状。整个建筑形式是在灵活多变中寻求秩序与逻辑，在风格上，这一建筑群是欧陆风情与中国传统建筑形式相结合的典范。

欧风街集餐饮、休闲娱乐为一体。地下一层设两个六级人员掩蔽所和一个人防物资库及公用设备站房，如配变电所、通风机房、消防及生活水池、移动发电机房等。人防用房平时可用作停车库，战时做必要封堵为人防掩蔽所。车库停车位337辆属Ⅰ类车库，车库设两个双向出入车道，地上二层为餐饮休闲娱乐。

FEATURE | 专题

Profile

Located at central area of Xixin New Area, Zhang Jia Gang, the site is just 2 km away from the downtown and on the west to GuotaiRoad. The park is in the north end of the site, on the south to Dazhai River. It is not only a beautiful park, but a large commercial complex that integrates shopping mall, catering, leisure, entertainment, culture, residential and office. It puts forward a new idea from all these various aspects.

Planning and Architectural Design

Besides the pool, greening and leisure chairs in the courtyard, there are shrubs, flowers and lawn outside, forming a graceful and colorful shopping environment.

European-style Street was built up by ten irregular monomers, thick and closed external walls shape an old castle, three inward courtyards decorated with pool, greening and leisure chairs which link with information management center, cultural park and air corridor in the hills. The whole street is elegant outside and bustling inside. Irregular part of the

plane toward the inner courtyard is neatly cut out, constitute the atrium space, where the facade is constituted by the delicate wooden frame curtain wall, shaping a contrasts with the external brick wall. The multi-fold sloping roof is in Chinese traditional style. The plan is irregular, so are the facade and the roof. The entire architectural form is seeking order and logic in a flexible style building complex, and is a model of European style and Chinese traditional architectural form.

European-style Street set catering, leisure and entertainment. Two shelters for six-level air-raid, materials storage and public equipment room, such as substation, ventilation equipment room, fire control and drinking water tank and mobile generation were set in the basement. The civil air defense space usually can be used as a parking garage or as a shelter in the wartime. 337 garage parking spaces is a class I garage, where there two double-way access lanes are equipped. Catering and entertainment are arranged in the first floor.

DELICATE AND RESTRAINED RESIDENCE IN SPANISH SOUTHERN CALIFORNIA STYLE | Tianjin Carmel

精致内敛的西班牙南加州风格住区
—— 天津卡梅尔

项目地点：中国天津市
业　　主：天津宁发发展有限公司
建筑设计：上海海波建筑设计事务所
合作设计：天津市房屋鉴定勘测设计院
摄 影 师：胡义杰
面　　积：151 787 m²
容 积 率：1.48
绿 地 率：45.1%

Location: Tianjin, China
Client: Tianjin Ningfa Group Co., Ltd.
Architectural Design: Shanghai Haibo Architectural Design Co., Ltd.
Collaborators: Tianjin Real Estate Appraise Survey & Design Institute
Photography: Hu Yijie
Area: 151,787 m²
Plot Ratio: 1.48
Greening Ratio: 45.1%

总平面图 Site Plan

项目概况与规划布局

　　卡梅尔位于天津梅江南7号8号地块。规划总用地面积为102 700 m²，建筑形态分为别墅、花园洋房、高层住宅及商业。此次规划设计为其中高层住宅部分，总建筑面积约151 800 m²，由8栋30层的高层住宅组成。

NEW CHARACTERISTICS | 新特色

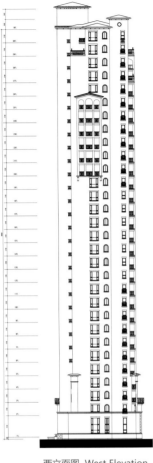
西立面图 West Elevation

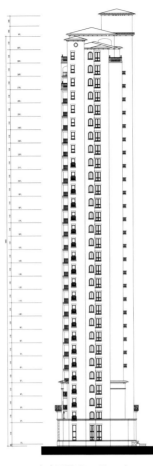
东立面图 East Elevation

建筑设计

建筑外立面设计选择了西班牙南加州风格，借鉴了很多西班牙别墅区的建筑符号。底层连续的拱廊、文化石墙体基座，上部每户宽大的透空柱廊阳台、木制隔栅，以及穿插的西班牙筒瓦小屋檐和各色装饰线条，构成强烈的西班牙式建筑风格。与项目中西班牙别墅区风格统一，完美融合，同时也为在高层住宅中体现西班牙风格的探索提供了新的思路。

户型设计

项目中7号地块4栋为两梯两户，房型面积是190~260 m^2大户型；8号地块4栋为两梯三户，房型面积是110~170 m^2中小户型。

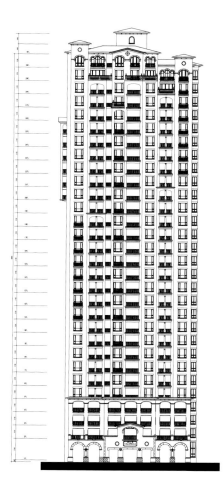

南立面图 South Elevation

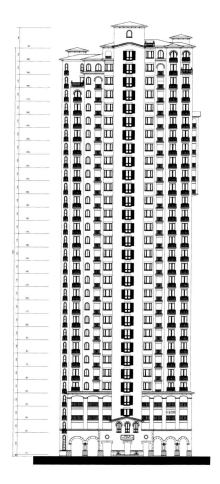

北立面图 North Elevation

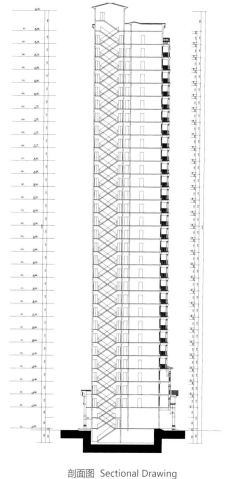

剖面图 Sectional Drawing

NEW CHARACTERISTICS | 新特色

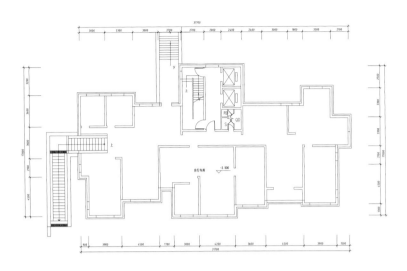

5# 地下层平面图 5# Basement Plan

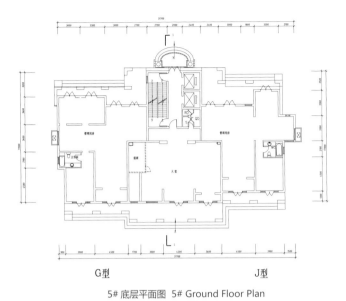

G型　　　　　　　　　　　　　　J型

5# 底层平面图 5# Ground Floor Plan

G型　　　　　M型　　　　　J型

5# 二层平面图　5# Second Floor Plan

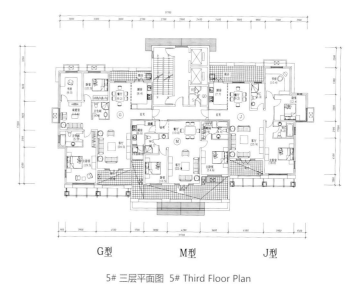

G型　　　　　M型　　　　　J型

5# 三层平面图　5# Third Floor Plan

NEW CHARACTERISTICS | 新特色

Profile and Planning
Tianjin Carmel locates in Land 8, Meijiangnan No.7 of Tianjin. It covers a land area of 102,700 m^2, used as villa, garden house, high-rise residence and commercial land. The project mainly develops medium and high-rise residential buildings with a total of 151,800 m^2, including 8 separate 30-story residential towers.

Architectural Design
The facade design chooses the Spanish Southern California style and borrows many architectural symbols from the Spanish villas. The sequential arcades & cultural stone wall base on the ground floor, large transparent colonnade balconies & wood retainer on the upper floors as well as the alternate Spanish imbrex eaves & all kinds of decorative mounding make up a strong Spanish architectural style. Thus realizing the agreement with the existed Spanish villas and also providing a new idea for studying the Spanish architectural style in the high-rise residential buildings.

Housing Type Design
Land 7 of the project provides 4 buildings of large household varying from 190 m^2 to 260 m^2. And Land 8 has 4 buildings with houses in small or medium varying from 110 m^2 to 170 m^2.

NEW SPACE | 新空间

BUILDING A MODERN SPACE WITH "GERMAN ELABORATION"

| ICON.Dayuan International Centre Sales Department
缔造"德系精工"品质现代空间——ICON•大源国际中心售楼部

项目地点：中国四川省成都市
室内设计：多维设计事务所
设 计 师：张晓莹
面　　积：800 m²

Location: Sichuan, China
Interior Design: Dodov Design Studio
Designer: Zhang Xiaoying
Area: 800 m²

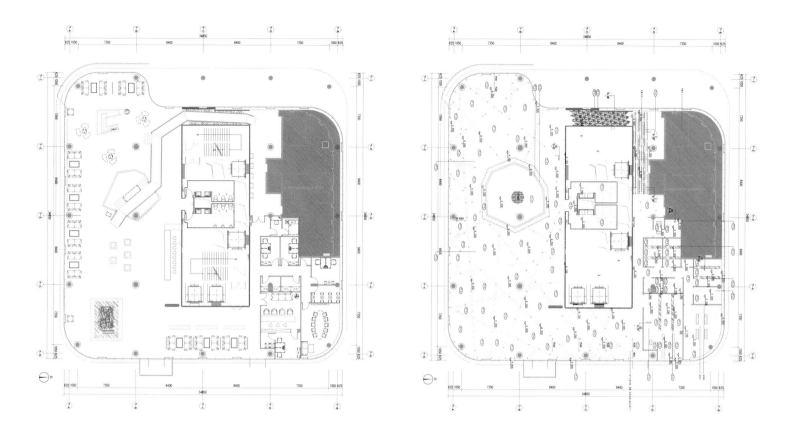

NEW SPACE | 新空间

项目位于天府大道旁大源CBD核心商务区，为2011年高投置业城南点睛之笔。外观采用德国GMP"德系精工"建筑手笔。

设计上结合项目和目标客户群特征，定义ICON•大源国际中心售楼部为现代风格，理念定位为"德国精工再发现"，项目logo演变的无规则现代感的线条、镀膜玻璃折面异形"钻石"盒子、德国精工品质的收藏品、硬朗现代感的家居等设计手法，呼应德国品质坚实的外表同时打造具有理性化、个性化、可靠化、功能化的内在空间特征。

灯光布置上，采用泛光源、点光源，LED线光源相结合，天花T5灯槽纵横交错，造型来源于项目logo，简洁明了，干脆利落。德国精工展示区域采用LED线光源，用透明亚克力为传送媒介。

Located in Dayuan CBD besides near with Tianfu Avenue, the building is the highlight that Goht Investment placed in the south of this city. Its exterior is with architectural technique of "German Elaboration" that meeting German GMP requirements.

Designer, combining the features of this project and targeted customers, defines the sales department of Dayuan International Centre as modern style and positions the concept as "the rediscovery of German Elaboration". Evolving from project's logo, the design technique, like irregular lines with stylish modern, irregular shape "diamond" box made from coated glass converted face, collections with "German Elaboration" quality and home furnishing of tough modern, corresponds with firm surface. In the meanwhile, such technique also creates an inner space with features of rationalize, individuation, reliability and functional.

As for disposition of lights, designer's inspiration is from project's logo which is simple and clear. So he combines flood light source, spot light source and LED light source. And troffers are crisscrossed. And in the area of "Germany Elaboration" displays uses, LED light source is used and transparent acrylic is conveying medium.

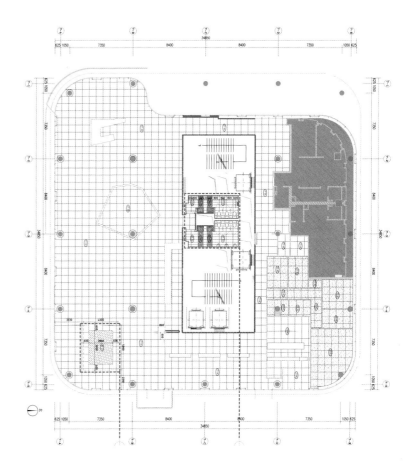

WATERFRONT RESIDENCE WITH DISTINCTIVE MOSAIC FACADE

| Apartment Building on the Harbour

独特"马赛克"图案立面的滨水居所—— 海港公寓大楼

项目地点：意大利拉文纳
客　　户：Iter Cooperativa Ravennate Interventi Sul Territorio
建筑设计：意大利Zucchi建筑师事务所
设计团队：Cino Zucchi, Nicola Bianchi, Andrea Viganò with Leonardo Berretti, Ivan Bernardini, Juarez Corso, Marcello Felicori, Chiara Toscani
规　　模：4 754 m²
摄　　影：Cino Zucchi

Location: Ravenna, Italy
Client: Iter Cooperativa Ravennate Interventi Sul Territorio
Architects: Zucchi & Partners
Design Team: Cino Zucchi, Nicola Bianchi, Andrea Viganò with Leonardo Berretti, Ivan Bernardini, Juarez Corso, Marcello Felicori, Chiara Toscani
Dimensions: 4,754 m²
Photography: Cino Zucchi

项目概况

　　这栋新建的住宅楼是大型城市重建项目中的一部分，紧邻拉文纳车站，位于一条人工运河的两侧，在该地工业区起着港口的作用。重建项目的整体规划由Boeri工作室设计而成，设计师们希望建立一座新公园，这座公园与水面和一系列河流两岸的高大建筑相平行，目前运河两岸还属于港口专用，但是一段时间之后会对公众开放。由于未来的场地用途尚未确定，因而设计师们为这栋住宅大楼打造了双面外观，既与现有的城市建筑风格紧密相连，同时又准备向水边开放，为将来可能转变成一个漫步场所做准备。在面向城市的一侧，绿色的土墙勾勒出室内停车场的轮廓，同时形成了一个向上隆起的中央庭院，可从这里俯瞰运河。通往电梯竖井的许多小型商店以及中庭，面向着这个半开放的"广场"，一条与建筑相平行的斜坡将之与滨水漫步道紧紧相连。

建筑设计

　　大楼分为两个体量，其几何形式各不相同，在中间衔接它们的可居住的"桥"面对着运河一侧，这种结构为中庭部分营造了一种空间上的闭合感和亲密感。两栋楼的不同高度既未阻碍眺望市中心的视野，也考虑到了这个综合建筑物的朝阳。北立面的设计使建筑看上去相当"庞大"，而南立面则将悬挑阳台的长水平线条当成特色。建筑物的主要立面是用一系列"有凹槽的"赤褐色水平模板装饰的，每一层立面使用两排模板，从而构成了一种彩色的灰泥抹面，即各种温暖的黏土色与钴蓝色的模板构成了"马赛克"图案——这是受拉文纳著名的拜占庭艺术风格的启发——让人们看到建筑物时容易产生一种图案失真的错觉。这种在建筑物规模上的瞬间"伪装"效果，有助于将实际的建筑规模与从水边和城市中对建筑的感知规模联系在一起。这栋大楼单独屹立在运河边，成为一个临时的标志性建筑，静静地等待着周围景观与建筑物的进一步改造。

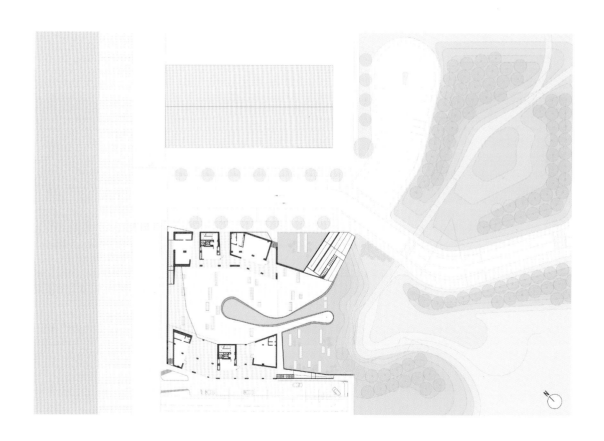

NEW IDEA | 新创意

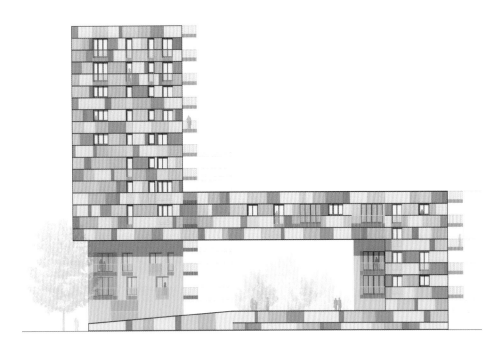

整栋建筑是按照以下"可持续发展"和节能建筑的最新标准进行构想的，并且实现了这些标准。设计师们仔细研究了建筑物主体与太阳方向之间的关系，并且针对建筑物的表面和开放的公共空间深入研究了所有时间和季节中的阴影图案。两个体量中较高的一栋位于北侧，较低的则位于南侧。建筑物南侧的几排阳台为起居室遮挡住了夏季的阳光，同时使得角度较小的冬季阳光能够照射到室内，极大地提高了建筑物的能源效率。较小的洞口是北立面的特色所在，这些洞口降低了热透射率。建筑物所需能源的重要组成部分是由安放在两栋建筑物屋顶平台上的太阳能电池板提供的。在钢筋网上涂抹一层灰泥，在建筑表皮形成了厚厚的保温隔热材料，这使得热透射率非常低，既节省能源，同时也为公寓营造了极为舒适的环境。所有的材料都能进行生物降解，易于处理，这些材料包括：石质的窗台、铁衫木的窗框以及表面所用的天然灰泥。

NEW IDEA | 新创意

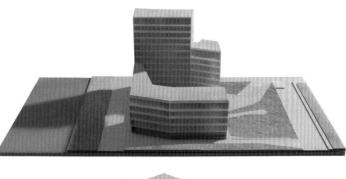

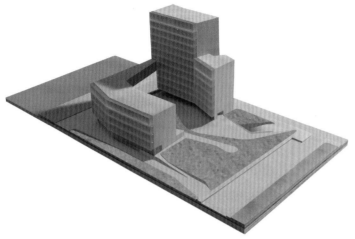

Profile

The new residential building is part of a large urban renewal project next to the Ravenna station on both sides of an artificial canal serving as a harbour for the industries of the area. The overall plan by Boeri studio envisages a new park parallel to the water and a series of rather tall volumes along the waterfront, which currently belongs to the harbour precinct but should in time become open to the public. Within the uncertainty of this "Sliding Doors" future, we designed a double-faced residential complex, relating to the existing city fabric but ready to open toward the water edge and its possible future transformation into a promenade. On the city side, a green rampart hosting the covered parking leads to a raised central court overlooking the water. A number of small shops and the atriums leading to the vertical distribution shafts open onto this semi-public "piazza", which will be connected to the water promenade by a ramp running parallel to the structure.

Architectural Design

The geometric inflections of the two building blocks and the lived-in "bridge" connecting them on the water side contribute to give a sense of spatial enclosure and intimacy to the central court. The

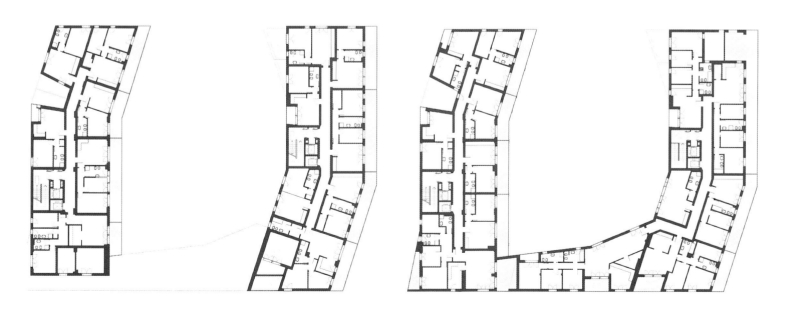

NEW IDEA | 新创意

different heights of the blocks are related to the long views toward the inner city and to the solar orientation of the complex. While the north side elevations of the buildings are treated in a rather "volumetric" way, the south ones are marked by the long horizontal lines of the overhanging balconies. The main facades of the building are marked by a number of terracotta horizontal "notched in" mouldings–two for every floor height–framing a plaster rendering of different shades of warm, clay-coloured shades in different hues and a cobalt blue one, creating a "mosaic" pattern–somehow inspired by Ravenna's famous Byzantine art–which generates a sort of scalar distortion in the perception of the building. This effect of momentary "camouflage" of the dimension of the building helps connecting its "domestic" dimension to its perception from the waterside and the city, where it stands alone as a temporary "landmark" waiting for the development to transform the landscape or this part of the city.

The whole complex is conceived and realized following the latest criteria for "sustainable" and energy conscious buildings. The building masses are carefully studied in relationship with the sun orientation, with an in-depth study of the shadow pattern at all hours and seasons both on the building surfaces and on the open collective spaces. The higher building is located on the north side and the lower on the south one. The rows of balconies on the south side of the buildings screen the living rooms from the summer sun rays, while admitting the lower winter ones, greatly contributing to the energy efficiency of the complex. The north facades are marked by smaller openings contributing to low thermal transmittance. A significant part of the energy required by the building is provided by solar panels placed on the rooftop terraces of the two buildings.

The thick "overcoat" insulation finished by a layer of plaster on mesh provides very low values of heat transmittance, saving energy and creating high environmental comfort for the dwellings. All materials are biodegradable or easily disposed of: stone for the window-sills, hemlock wood for the window frames, natural plaster for most of the exterior surfaces.

COMME
BUILDINGS

RCIAL 商业地产

P130 赛特71商住公寓：
历史与现代元素的碰撞融合

云龙数码科技城规划设计：
绿色生态 科技新城

P142 CIB生物医学研究中心：
仿生学原理在建筑中的综合运用

THE INTEGRATION OF PAST AND MODERN | 71 council and private flats in Sète

历史与现代元素的碰撞融合
—— 赛特71商住公寓

项目地点：法国赛特
客　　户：Pragma
建筑设计：法国Colboc Franzen建筑师事务所
设计团队：Benjamin Colboc, Manuela Franzen, Arnaud Sachet
可用面积：3 913 m²
摄　　影：Cécile Septet

Location: Sète, France
Client: Pragma
Architects: Colboc Franzen & associés
Project Team: Benjamin Colboc, Manuela Franzen, Arnaud Sachet
Usable Area: 3,913 m²
Photography: Cécile Septet

项目概况

项目位于托湖与古镇以南的地中海之间的一狭长地块中，靠近有着庞大工业设施的商业港口。设计师希望通过建筑再现历史的同时，赋予古镇门户及其周边新兴现代特性，在不改变古镇生活方式的前提下，调整好建筑与巨型港口的关系。

规划布局

项目有三栋楼，包括4个明显的部分：16套配置不同的公营公寓，55套私人两房、三房公寓，商店和停车场。商店和停车场位于底层。六层公营公寓位于项目正中，作为新旧建筑的过渡，其余两栋八层楼宇可自由地展示自身特点。位于街角的一栋，标志着古镇的入口，可远眺港口和即将开发的空置码头区。

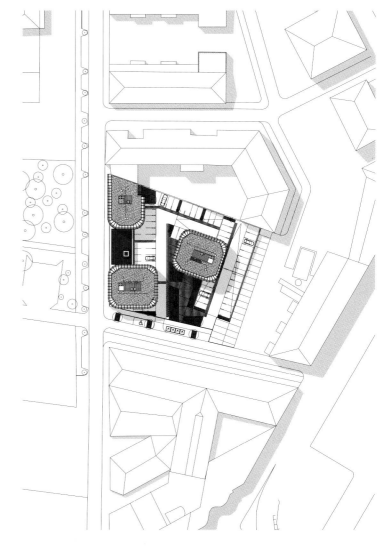

总平面图　Site Plan

COMMERCIAL BUILDINGS | 商业地产

建筑设计

项目遵循"地中海建筑"的设计原则，以便让住户享有适应当地气候的生活方式：拒绝酷热，尽享户外空间。建筑外立面由露台围合，住户可随意走动。外层的镀锌钢板屏缓解了酷热，也保证了公寓的私密性。它们就像巨型的钢蚕茧，让人联想起海洋世界：形如船首，穿屏之风像敲打着桅杆一样，发出清脆的叮当声。这些钢板屏也有利于住户合理运用露台空间、邻里间互不侵扰，营造家的感觉，让用户欣赏美景、享受古镇生活。

景观设计

项目充分利用坡地地形，将景观花园与停车空间置于地块中心。灌木和小乔木带来一片阴凉，对地表起到保护作用，也拉开了公寓与停车场的必要距离。

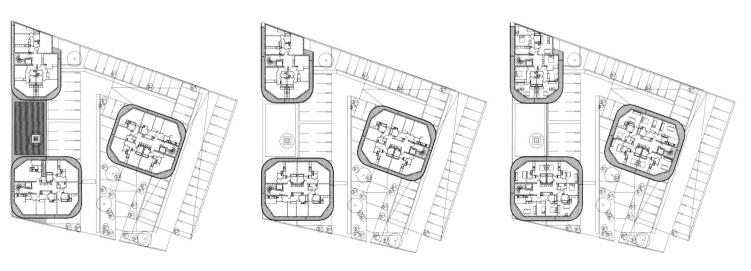

COMMERCIAL BUILDINGS | 商业地产

Profile

The building plot lies on the thin strip of land between the Étang de Thau and the Mediterranean Sea on the northern side of the old town, close to the commercial port and its huge industrial facilities. How should the designers evoke the site's past and at the same time, through architecture, forge a modern identity for this entrance point to the town of Sète and its emerging neighbourhoods? How should they respond to the titanic scale of the port, with the sea as the horizon, while also maintaining the old town's way of living?

Planning and Layout

The project design is based on three blocks of flats set on a ground-floor base. The development comprises four distinct parts: 16 council flats in various configurations; 55 private two- and three-room flats; and shops and car parks to service all of the above. The base accommodates the shops and the car parks, whereas the blocks house the flats. The six-floor block of council flats provides a transition from the existing buildings around it and is therefore located at the centre of the project. The other two eight-floor blocks are thus free to demonstrate their autonomy. The block standing on the street corner marks the entrance to the old town while also looking out towards the commercial port facilities and future developments on the empty docklands. The block at the back is situated above parking spaces and gardens. It looks like a sculpted object in the middle of the 'island' and we therefore forget that regulations made it impossible to set the building against the existing party wall.

Architectural Design

These blocks also embody a principle of 'Mediterranean architecture' that allows for a lifestyle adapted to the local climate: outdoor living protected from intense heat. There are balconies running along the facade and these outdoor extensions allow occupants to walk around the outside of their flats. A galvanised steel screen protects it during very hot weather and also provides a nice amount of privacy. It follows the curve created by the varying widths of the balconies. It lends harmony to the three blocks and makes them easier to interpret. They become gigantic steel cocoons whose materials remind us of the maritime world, while their shape is reminiscent of a ship's stem and the wind in the screen slats sounds like the jangling of masts in a port. The screen also allow occupants to make appropriate their balconies without disturbing their neighbours, and to create a 'homely' feel while also enjoying the view and life in the town centre.

Landscape Design

Making good use of the various slopes, the car park creates a man-made topography in the centre of the block of land and harbours a landscape of gardens and parking spaces. The effect is like shelly limestone and it is punctuated with beds of broken rocks and characteristic regional plants. The ground is protected by a layer of bushes and small trees, which provide shade as well as establishing the requisite distance between the flats and people using the car parks.

GREEN AND ECOLOGICAL, NEW TECHNOLOGY CITY

| The Planning and Design of Yunlong Digital Technology City

绿色生态 科技新城 —— 云龙数码科技城规划设计

项目地点：中国湖南省株洲市
开 发 商：湖南中京泰安实业有限公司
建筑设计：深圳市佰邦建筑设计顾问有限公司
　　　　　NAUTA architecture & research（荷兰）
总用地面积：480 000 m²
总建筑面积：936 000 m²
容 积 率：2.0
绿 地 率：40.5%

Location: Zhuzhou, Hunan, China
Developer: Hunan Zhongjin Taian Industrial Co., Ltd.
Architectural Design: P.B.A Architectural Designing Ltd.
　　　　　　　　　　NAUTA architecture & research
Total Land Area: 480,000 m²
Total Floor Area: 936,000 m²
Plot Ratio: 2.0
Greening Ratio: 40.5%

项目概况

云龙数码科技城位于中国湖南省长株潭城市群云龙示范区内。具体选址于长株高速以东，云龙大道以南，云海大道以北的菖塘村板块。园区规划用地总面积480 000 m²。

为适应长株潭城市群一体化发展的需要，并积极建立国家两型社会改革实验的示范工程，本项目着眼于探索可持续发展的新型科技园发展模式，打造资源节约型与环境友好型的产业地产，实现产业城资产的长期增值，从而带来巨大的经济效益，社会效益与环境效益。

规划布局

总体规划布局充分考虑了与各规划道路之间的关系，力求将东面开发为城市绿地，南面为园区内的公共绿地，这样的布局将城市空间与园区景观进行了有机的融合。方案坚持建筑产品多样化、组团空间多元化、设计人性化，着重发展低

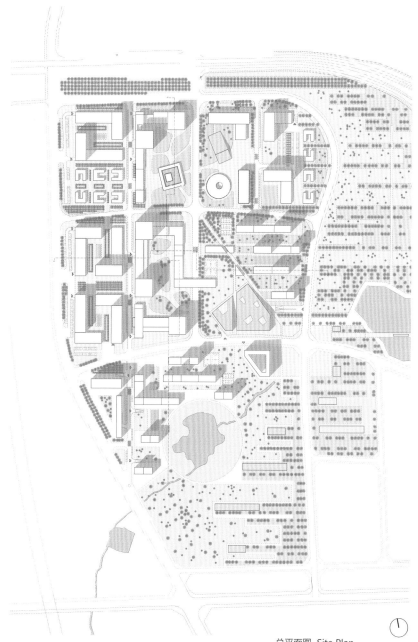

总平面图 Site Plan

COMMERCIAL BUILDINGS | 商业地产

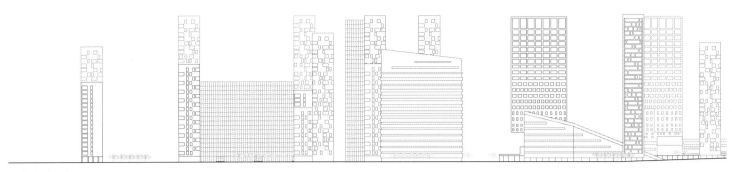

沿湖东立面
east facade

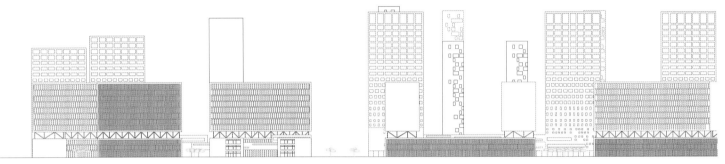

沿高速西立面
west facade

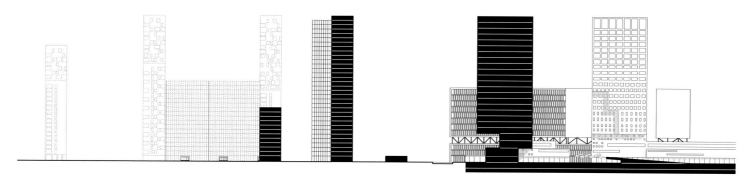

整体剖面AA
section AA

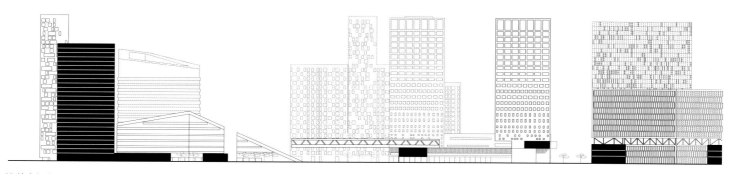

整体剖面BB
section BB

碳、生态、绿色设计，力争打造成为高生态含量、人与自然和谐共处，经济社会与资源环境协调发展的科技新城。

建筑设计

建筑设计方面，项目整体体型方正、大气，在设计的过程中，除了注重轻盈、整体感，同时也注意其通透性；建筑立面形式极其丰富，通过肌理渐变、间距疏密控制等处理手法，淋漓尽致地体现了园区的时尚、年轻、活力、国际化等品质。

景观设计

项目景观设计注重其功能性、生态性、经济性，满足市民休闲、娱乐、游览的需求；强调公园在城市生态系统中的作用，强调人与自然的共生；同时充分利用场地条件，减少工程量，考虑景观的经济效益。

近40.5%的绿化率让每一个人都能置身于园林风景中。景观设计采用休闲、生态、实用为一体的都市现代自然主义风格，带来"亲、静、美、闲、雅、秀"的温馨享受。

COMMERCIAL BUILDINGS | 商业地产

首层总平面图 1/4000
GROUNDFLOOR MASTER PLAN 1/4000

Profile

The project locates in the Yunlong New Town of Changsha-Zhuzhou-Xiangtan (CZT) Region, Hunan Province, China. It is located in the Changtangcun area that is in the east of Changzhu regional highway, in the south of Yunlong Road and in the north of Yunhai Road. The park covers a total site area of approximately 48 ha.

This project is driven by the intention to integrate the cities of Changsha, Zhuzhou and Xiangtan into one urban region. Furthermore it should respect the National vision that aims to develop this region as a testing zone of 'Resoucesaving and Environmentally friendly development'. Therefore, this project has the ambition to explore a sustainable model to develop a business park, which could guarantee a long-term payback for the investors in terms of economy, social and environmental development.

Planning and Layout

The project fully takes the road planning into consideration and tries to develop its eastern part into the urban green space and western part into the public green land inside the park to achieve the reasonable integration between the urban space and the park. This project pursues for product diversification, group space diversification, design humanization and the low-carbon eco green design so as to develop this area into a new technology city which is highly ecological, harmonious and achieves the harmonious development between the social economy and resources environment.

Architectural Design

The project goes for upright and generous design, focusing on the qualities of lightness, wholeness and thoroughness. The gradual texture change and separation distance control contribute to the abundant facade forms and fully show the fashion, juvenility, vigor and internationalization of the park.

Landscape Design

The landscape design lays emphasis on the functional, ecological and economical efficiency to meet the citizens' demand of relaxation, entertainment and touring. It concerns the effect of the parks in the urban ecosystem and the mutualism between the nature and man. It proposes to take full advantage of the site condition to lessen the project amount and realize the economic benefit brought by the landscape design.

The greening ratio of 40.5% makes this area a garden scenic spot for everyone. The design applies the urban modern naturalism style that blends the qualities of relaxation, eco and practical to bring people a peaceful and elegant living environment.

IMPLEMENTATION OF BIOMIMICRY IN ARCHITECTURE

CIB – BIOMEDICAL RESEARCH CENTER

仿生学原理在建筑中的综合运用 —— CIB生物医学研究中心

项目地点：西班牙潘普洛纳
建筑设计：Vaillo & Irigaray & Galar建筑师事务所
占地面积：12 150 m²
摄　　影：Jose Manuel Cutillas、Pedro Pegenaute 和 Ruben P. Bescos

Location: Pamplona, Spain
Architects: VAÍLLO & IRIGARAY & GALAR
Surface: 12,150 m²
Photography: Jose Manuel Cutillas, Pedro Pegenaute and Ruben P. Bescos

项目概况

通过在建筑过程中应用仿生学原理，骆驼、北极熊和树叶这三种互不相干的事物在这座研究中心上得到了前所未有的综合运用。建筑师利用这三种生物原型创造了一个类似的自适应系统。

建筑设计

如同骆驼的身体结构能够根据身体需要进行扩展一样，屋顶上不同尺寸的突起以及地标和半地下的中空结构也是为了满足建筑功能的需要。因此，为了提高建筑的工作效率，也为了适应未来的变形需要，在不丧失建筑本质的前提下在设计之初就对建筑整体进行的变形。

"北极熊"的皮肤可以保持熊体内的温度抵抗外界的严寒，与此相仿，这座研究中心根据这一原理使用黑色的厚表皮结构，包裹住内部有孔的透明"毛发"以达到保持内部低温的目的。设计师采用的白色的"毛发"不仅出于建筑经验的积累，同时还可以与周围的冰雪世界相映成趣。

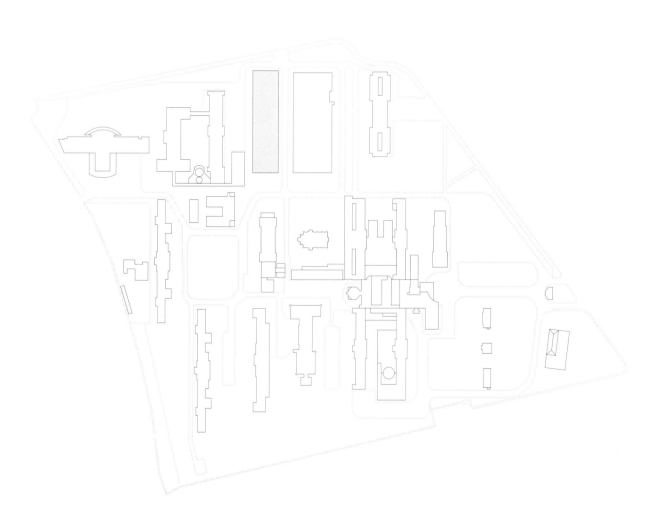

COMMERCIAL BUILDINGS | 商业地产

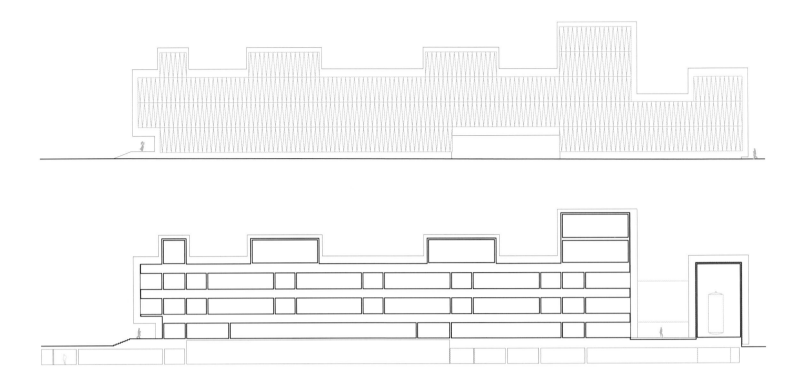

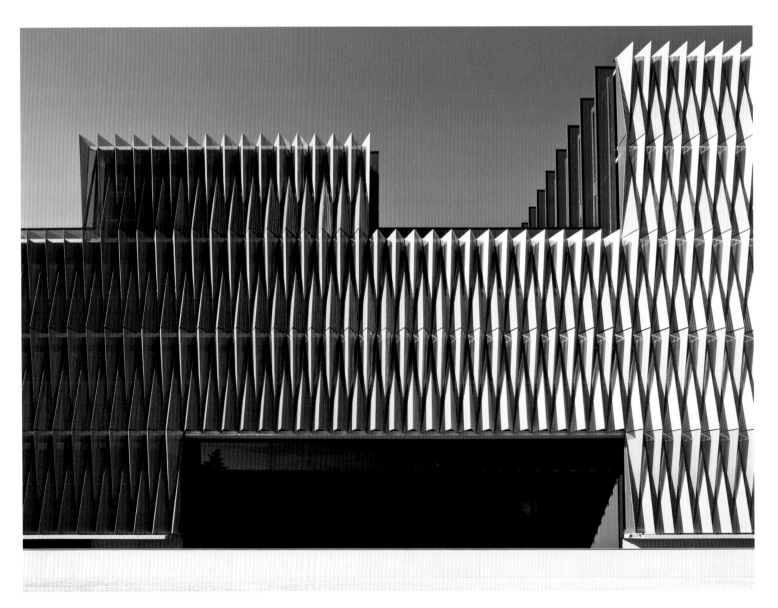

研究中心使用的串孔钢板设计引入了第三种生物原型——树叶：这些面积巨大而厚度减薄的钢板需要自行支撑重量，需要有很强的弹性并保持垂直状态。折线的交错如同树叶表面的脉络，为整体提供更强的刚性并达到要求的几何结构，为建筑进行隔热同时还扩大了视野。

由Vaillo & Irigaray & Galar建筑事务所设计的CIB生物研究中心呈现出的形象与其功能相适应，通过一个统一的表层形式将建筑的特色表达出来。而且，这个外层的表皮将整个内部的结构包裹起来，并赋予项目更强的功能差异性。

Profile

The camel, the polar bear and the leaf: the project aims to link with the content of the program: Bio-Medical Research, through the implementation of the biomimicry in the process of generation architecture. Architects take these three bio-reference types to achieve similar adaptive systems

Architectural Design

As well as the camel's anatomy expands towards his function requires, in this building bumps and hollows are generated according to requirement of the function: with the creation of backpacks with different sizes on the roof and hollows on the ground floor and semi basement. So, the building deforms to configurate a silhouette able to operate effectively, even being able to mutate for the future on its deformations, without loosing its essence.

The skin must keep a consistent inner temperature of the bear, despite the permanent outdoor cold, and it achieves it trough a black thick skin, wrapped in holed transparent hair that keeps the air cold inside. The vision we have is of white hair, but it is due to the accumulation, and so it mimes with ice and snow.

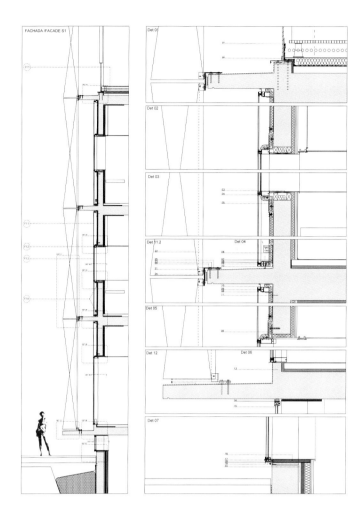

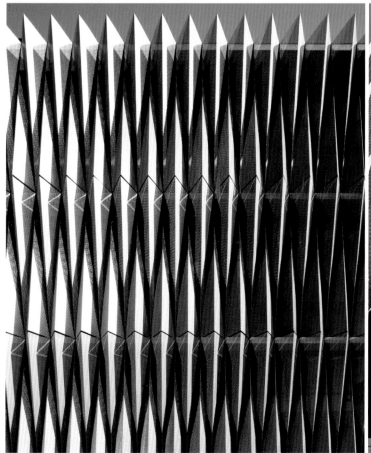

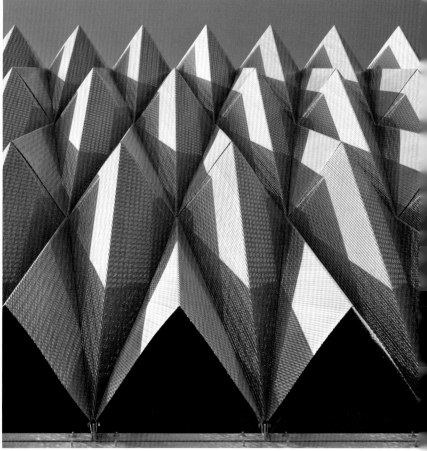

COMMERCIAL BUILDINGS | 商业地产

For the design of the perforate steel plates, the third biotype is involved: As well as the leaves of the trees, those huge plates (4500x800mm) and with reduced thickness (3mm) have to be self-supporting, light, flexible and keep upright…to that end the mixture of the origami allows the generating of plans, folds and nerves, providing rigidity to the set and taking the required geometry to protecting from sun and allowing the vision.

The building proposes an image inherent in its intrinsic functionality and therefore is manifested by an envelope that covers their characteristic forms. Somehow, the outer skin "layers" their internal structures. Plants are extreme, which encompass parts of the program more functional disparity.

生态 | 节能 | 低碳

从理想到现实，从卓越到经典

奥森国际景观规划设计荣获

2012年IDEA-KING艾景奖

2012年优秀景观设计机构

颁发机构:国际园林景观规划设计行业协会,世界屋顶绿化协会

奥森国际设计宗旨是从景观的视觉价值、文化价值、体验价值、生枋价值基点出发，为客户和终端客户创造增值空间；创造具有艺术感力和科学性的景观设计；创造有生命的景观，达到人与自然和谐统一并致力于在这一过程中成为行业的榜样！

禹州高尔夫项目实景照片

城市规划	住宅区景观	商务空间	市政公园
Planning and Urban Design	Residential area landscape	business space	mulniciple park

奥森国际景观规划设计有限公司

OCEANICA INTERNATIONAL LANDSCAPE PLANNING DESIGN.,LTD.

奥森国际景观（Oceanica International landscape planning design）于2003年在中国注册成立，拥有一批来自世界各地具有丰富实践经验的景观设计师，核心专业团队在国内外一线景观设计机构均具有十年以上的工作经验。现拥有主要设计成员约50多人，项目遍布全国各省市，曾为多家知名地产商服务过，包括万科、万达、金科、保利、恒大、中信、中铁、南国奥园、佳兆业集团等。公司凭借"专业""经验丰富"和"服务到位"，获得越来越多得到社会的关注，得到各方面专家和客户的高度评价。

| 奥森国际景观规划设计有限公司
OCEANICA INTERNATIONAL LANDSCAPE PLANNING DESIGN.,LTD. | www.oc-la.com | 【地址】深圳市南山区南海大道粤海路动漫园7栋5层
【邮编】518000【电话】0755-26828246【传真】0755-26822543 |

四季园林

鸿艺集团．客天下．瀑布教堂

■ 风景园林专项设计乙级
- 景观设计　　　　Landscape Design
- 旅游建筑设计　　Tour Architectural Design
- 旅游度假区规划　Resorts and Leisure Planning
- 市政公园规划　　Park and Green Space Planning

广州市四季园林设计工程有限公司成立于2002年，公司由创始初期从事景观设计，已发展为旅游区规划、度假区规划、度假酒店、旅游建筑、市政公园规划等多类型设计的综合性景观公司。设计与实践相结合，形成了专业的团队和服务机构，诚邀各专业人士加盟。

Add：　广州市天河区龙怡路117号银汇大厦2505
Tell：　020-38273170　　　　Fax：　020-86682658
E-mail：yuangreen@163.com　　Http：//WWW.gzsiji.com

寻找中国房地产真英雄
CIHAF 2012
中国房地产"三名"（名人·名企·名盘）大奖
———— 记录中国地产光荣与梦想 ————

2012年12月6日 / 中国深圳

奖项设置

 CIHAF 2012 中国房地产名人推介榜
中国房地产年度影响力人物

 CIHAF 2012 中国房地产名企推介榜
涵盖房地产行业具有品牌影响力、行业整合力、社会公益推动力、建筑科技引导力、行业创新力等优秀代表企业。

 CIHAF 2012 中国名盘推介榜
代表2012年中国房地产年度开发水平，具有行业标杆意义的房地产项目。

 CIHAF 2012 特别奖项推介榜
包括绿色建筑、商业地产、产业地产、旅游地产、养老地产以及城市规划、新闻媒体等领域。

评审流程

6月-9月
企业自愿申报、各地主流媒体及相关机构推荐

10月
CIHAF中国住交会官方网站 www.cihaf.cn
公示候选企业、项目、人物参评资料。

11月
召开推介委员会总体评审会议，评议候选企业、项目、人物参评资料。

12月
CIHAF中国房地产"三名"大奖颁奖盛典

———— **CIHAF中国房地产主流媒体联盟** ————

全国六十余家主流媒体　　一年一度CIHAF强力整合传播

网上申报通道　　010-65085603
www.cihaf.cn　　010-65079988